The Yuma Reclamation Project

The Yuma Reclamation Project

Irrigation, Indian Allotment,
and Settlement Along the
Lower Colorado River

Robert A. Sauder

University of Nevada Press
RENO & LAS VEGAS

University of Nevada Press, Reno, Nevada 89557 USA
Manufactured in the United States of America
Design by Kathleen Szawiola

Library of Congress Cataloging-in-Publication Data
Sauder, Robert A.
The Yuma reclamation project : irrigation, Indian allotment, and settlement along the
lower Colorado River / Robert Alden Sauder.
p. cm.
Includes bibliographical references and index.
ISBN 978-0-87417-783-1 (hardcover : alk. paper)
1. Reclamation of land—Arizona—Yuma Valley—History—20th century. 2. Irrigation—
Arizona—Yuma Valley—History—20th century. 3. Indian allotments—Arizona—Yuma Valley—
History—20th century. 4. Indians of North America—Arizona—Yuma Valley—History—20th
century. 5. Land settlement—Arizona—Yuma Valley—History—20th century. 6. Yuma Valley
(Ariz.)—History—20th century. 7. Colorado River Valley (Colo.-Mexico)—History—20th century.
8. Yuma Valley (Ariz.)—Environmental conditions 9. Colorado River Valley (Colo.-Mexico)—
Environmental conditions. 10. Irrigation—Political aspects—West (U.S.)—History—20th century.
I. Title.
TC824.A6S25   2009
333.73'15309791—DC22   2009015671

FIRST EDITION
18 17 16 15 14 13 12 11 10 09
5 4 3 2 1

For Rose

When I was a kid in geography class,
I was taught that water always flows
downhill. What I've learned since is
that water flows to money and power,
wherever they may be.

—Former Navaho tribal chairman Peterson Zah,
quoted in *American Indian Water
Rights and the Limits of Law*,
by Lloyd Burton

Contents

Illustrations and Tables

MAPS

TABLES

Preface

As a historical geographer, my initial interest in the colonization of the arid West is revealed in my book *The Lost Frontier: Water Diversion in the Growth and Destruction of Owens Valley Agriculture*, published in 1994. *The Lost Frontier* focuses on the manner in which irrigated agriculture became established in the Owens Valley, California. It illustrates how the cooperative efforts of rugged pioneer farmers, without government assistance, turned an isolated region in the intermountain West into an agricultural oasis. It also details how Los Angeles, an increasingly thirsty metropolis, bought up the valley's land and water rights, resulting in the demise of Owens Valley agriculture. After completing *The Lost Frontier*, I began to set my sights on what I refer to in this book as the *final frontier*, the remaining irrigable, but not so easily irrigated, lands scattered throughout the western United States. In contrast to my Owens Valley study, which examines how individual pioneers went about reclaiming arid lands, this study focuses on the efforts of the federal government, through passage of the Newlands Reclamation Act, to bring the remaining irrigable (including Indian) lands in the arid West under irrigation and settlement.

The Yuma Reclamation Project is the primary concern of this study. Laguna Dam, located just above Yuma, was the first of many dams to be built on the Colorado River by the Reclamation Service, now the Bureau of Reclamation. I selected the Yuma Reclamation Project for investigation because it illustrates the multiplicity of problems and challenges associated with national irrigation policy in its attempt to facilitate homesteading in the arid West. Although the Yuma Project, because of its complexity of problems, might be considered exceptional among federal Reclamation projects, it nevertheless demonstrates the capacity of local case studies to enhance our understanding of the settlement processes that characterize the broader region in which they are situated. This study, then, will explore the interplay among and the consequences resulting from the multilayered federal policies designed to advance irrigation farming in the arid West, policies that met head-on in the vicinity of Yuma early in the twentieth century. Yuma, however, not only exemplifies the range of difficulties associated with settling the nation's final frontier but also helps us understand the source of some of the current issues and conflicts concerning the Colorado River.

The chapters that follow are organized more or less chronologically, with each chapter focusing on an important theme. The first chapter establishes the framework for this study by providing a brief review of the literature dealing with water resource development in the American West, and background information relating to the Dawes General Allotment Act, indigenous versus industrial irrigation, and the events that led to the passage of the Newlands Reclamation Act.

Chapter 2 outlines the territory originally dominated by the Quechans and describes the symbiotic relationship that developed between the tribe and the Colorado River. The natural setting of the lower Colorado River region is described, and the Indians' reliance on flood recession farming based on the river's overflow is reviewed. The impact of the Spaniards and subsequent intruders in Quechan land is then examined. The increased pressures of permanent settlers on land occupied by the tribe persuaded the federal government to place more precise limits on Quechan land—hence the establishment of the Fort Yuma Indian Reservation in the late nineteenth century. It will be shown how the tribe was reduced from a proud, self-reliant, and self-sufficient people to one dependent on government policies for survival, clearing the way for the irrigation frontier to advance into the region.

Chapter 3 discusses the evolution of irrigation near the Colorado River and Gila River confluence. The proposed plan of one irrigation enterprise led to an agreement with the Quechans allowing for the allotment of the irrigable portion of their reservation under the Dawes General Allotment Act. Even though this irrigation scheme failed to materialize, the seeds of Quechan allotment had been planted and would germinate a decade later. Meanwhile, irrigation and settlement on the opposite side of the river in the Yuma Valley were hampered by a sizable Mexican land grant and the difficulties of private irrigation companies to divert the mighty Colorado River to their newly planted fields.

The circumstances and complications connected with the approval of the Yuma Project and the onset of industrial irrigation, including the placement of the first dam on the Colorado River, are examined in chapter 4. The situation that prevailed along the lower Colorado River was metaphorical, for it reflected the federal government's concern that irrigation in the

West had gone about as far as it could in the context of individual effort and corporate enterprise. The engineering problems that had to be overcome on the lower Colorado River are discussed, as are the problems involved in the inclusion of the Fort Yuma Indian Reservation and the land in private ownership in the Yuma Valley. Attempts by the Reclamation Service to incorporate the Imperial Valley irrigation developments into the overall project are also examined.

Chapter 5 focuses on the developments that led to the allotment of the Fort Yuma Indian Reservation, even though nearly twenty years had elapsed since the approval of the Allotment Act and the assimilation of Indians into white society was no longer a priority of the federal government. The objective instead had shifted to the acquisition by non-Indians of potentially productive Indian lands. A number of issues surrounding the process of allotment are discussed, including the failure of Congress to inform the Quechans of legislation allotting their reservation or to seek their approval for the disposal of surplus lands, the difficulties of achieving an accurate count of the Quechans for allotment purposes, and the crusade to increase the size of Quechan allotments from five to ten acres.

Chapter 6 examines developments in the Bard unit of the Yuma Project, that portion of the Fort Yuma Indian Reservation designated as surplus land and opened to white homesteaders, and named after the senator whose untiring efforts made the Yuma Project a reality. The Bard unit represented that part of the project for which the Reclamation Service had especially high hopes, primarily because it was not encumbered by complications involving the Indian allotments, or settlers who were already on the ground, as in the Yuma Valley. But no screening of applicants for homesteads was done to determine their financial status or farming background, a process that represented a significant flaw in Reclamation Service settlement policy.

The array of problems associated with the delivery of water by the Reclamation Service to settlers in the Yuma Valley is the subject of chapter 7. Valley residents believed that federal intervention would bring to an end the difficulties that they had theretofore struggled with in their attempts to bring the Yuma Valley under irrigation. Their enthusiasm, however, was quickly dampened as an extraordinary series of problems befell the region in the early years of federal involvement, including major flooding on the lower Colorado, causing their crops to be inundated and their canals to be impaired, as well as delays in the completion of Laguna Dam and the Colorado River siphon, the link that connected them to the rest of the project.

Chapter 8 examines the multiplicity of new problems that arose on the project once water from the Colorado River began to be turned onto Yuma Project lands. It is apparent that federal policy makers and Reclamation Service engineers were oblivious to the many complex challenges associated with arid land reclamation, because, to their surprise, the turning of water onto dry land did not inevitably result in reclamation. Some of the unforeseen difficulties involved environmental repercussions resulting from the onset of industrial irrigation, while others were related to socioeconomic circumstances resulting from Reclamation policy. Nothing seemed to go right on the Yuma Project for the first decade after its opening. Given the naïveté of the Reclamation Service, combined with that of potential irrigators, it is a wonder that successful agriculture ever became established on Yuma Project lands.

Chapter 9 reveals how, in spite of the disappointments and conflicts connected with its early years of settlement and development, by 1920 the Yuma Project had turned the corner, and the agricultural possibilities so long promised were finally beginning to materialize. This chapter examines not only the gradual evolution of successful irrigation farming on the project but also some ongoing difficulties found there.

The final chapter examines the significance of developments at Yuma in the context of some of the many issues that have emerged since the river's transformation began more than a century ago with the construction of Laguna Dam. This early irrigation project led to subsequent developments on the Colorado that resulted in unforeseen troubles decades later. Hoover Dam and the All-American Canal, the Colorado River Compact, and the Mexican Water Treaty are discussed, as are problems concerning the quality of Colorado River water delivered to Mexico, California's attempts to live within its Colorado River allocation, and the ecological and economic impacts this may have on the Salton Sea and the Imperial Valley. It will be illustrated how the problems related to industrial irrigation on the lower Colorado River continue to multiply and become increasingly complex over time.

In the course of my investigation, I have traveled from coast to coast visiting various research centers. I have worked at the National Archives in Washington, D.C., and its Pacific Southwest regional branch in Southern California, at the Water Resources Center Archives and the Bancroft Library at the University of California–Berkeley, and at the Reclamation Bureau's Yuma Area Office in Yuma, Arizona, and I have examined relevant collections at Arizona State University, the University of Arizona, the Huntington Library, and elsewhere. Many individuals at the above institutions, the names of most

I do not know, provided me with invaluable assistance, without which this story could not have been told. I am deeply indebted to all.

Three persons deserve special recognition for their generous assistance. Kathy Carr, former public relations director of the Yuma Area Office of the Bureau of Reclamation, was instrumental in supplying me with many of the historical photographs that appear in this volume. Willa (Jeanie) Taliancich, formerly a geography student at the University of New Orleans (UNO) and currently an adjunct instructor at UNO and Tulane University, spent many hours creating publishable maps from my original hand drawings. Without the help of Kathy and Jeanie, this volume would contain considerably fewer aids to help the reader visualize my discussion. To my wife, Rose, who was by my side from the beginning of field and archival research to the final editing of the manuscript, I give my heartfelt thanks. Her untiring assistance, her keen eye, her thoughtful criticisms, and her incredible patience, all are reflected in the pages of this book.

This project began several years before my retirement from the Department of Geography at the University of New Orleans. UNO provided sabbatical support allowing me to embark on this study, and generous financial assistance through several Liberal Arts research grants to help finance my research. After my retirement and subsequent move to Tucson, the University of Arizona's Department of Geography and Regional Development offered me visiting scholar status in order to facilitate my ongoing research. I am deeply grateful for the privilege of being affiliated with both universities. I am also indebted to the anonymous reviewers who read earlier drafts of this study, and who offered constructive criticisms for improvement. Finally, I wish to thank the editorial staff of the University of Nevada Press, especially Charlotte Dihoff, who encouraged me to submit my manuscript and shepherded this project through the early stages of the publication process.

The Yuma Reclamation Project

1

The Final Frontier

On June 17, 1902, when Theodore Roosevelt signed the Newlands Reclamation Act into law, both Congress and the president had by then resolved that a national irrigation policy would be necessary in order to subdue the country's remaining irrigable, but not so easily irrigated, lands in the western United States.[1] Outside of Alaska, these lands, along with some Indian reservations that offered special agricultural possibilities coveted by non-Indians, would represent the country's final frontier, the last areas of the public domain where opportunities for homesteading by small family farmers could be developed (map 1.1). One such frontier was found along the lower Colorado River near Yuma, Arizona. Authorized in May 1904, the Yuma Reclamation Project, a by-product of the Newlands Reclamation Act, straddles the Arizona-California border (map 1.2). The Yuma Project originally encompassed ninety thousand acres of fertile Colorado River floodplain land, including the Fort Yuma Indian Reservation occupied by the Quechan (pronounced "kwuh-tsan") tribe. Yuma was one of the earliest federal irrigation projects to be initiated in the western United States, and was the first to be authorized on the Colorado River, a stream that drains much of the western United States (map 1.3).

Perhaps the most comprehensive overview of water resource development in the arid West is found in Donald Pisani's two-volume set titled *To Reclaim a Divided West: Water, Law, and Public Policy, 1848–1902* and *Water and the American Government: The Reclamation Bureau, National Water Policy, and the West, 1902–1935.* These works, which analyze nearly a century of western water policy, are the first two installments in an anticipated multivolume history of water policy in the American West. Another insightful discussion of water resource development in the American West is found in Donald Worster's *Rivers of Empire: Water, Aridity, and the Growth of the American West,* which demonstrates how today's West has evolved into a modern hydraulic society through the large-scale manipulation of water in the region. In a similar vein, but imbued with more

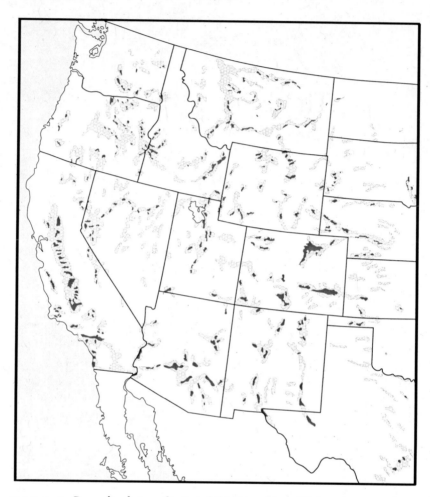

MAP 1.1. Generalized view of irrigated (*black*) and irrigable (*gray*) areas in the western United States, ca. 1900. (Based on Department of the Interior, U.S. Geological Survey, *First Annual Report of the Reclamation Service, From June 17 to December 1, 1902, 32*)

political intrigue, Marc Reisner's *Cadillac Desert: The American West and Its Disappearing Water* examines how water developers and government engineers have transformed the American West into a series of agricultural and metropolitan oases. Each of these works is an informative and thoroughly documented account of the political process and the consequences of western water development, but none focuses specifically on the process of settlement in the arid West. A truly comprehensive appreciation of the

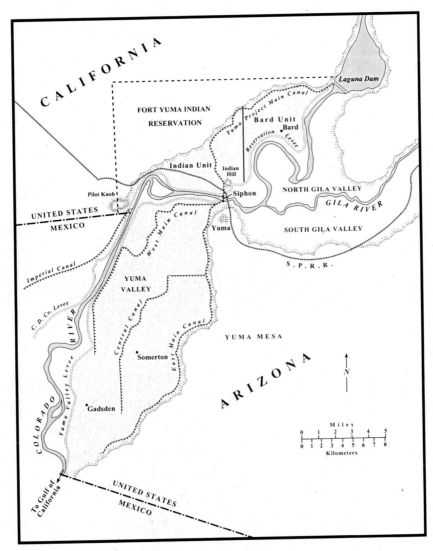

MAP 1.2. Yuma Reclamation Project.

American West's agricultural transformation awaits the completion of detailed case studies focusing on settlement, studies that can ultimately be combined into a mosaic where common strands linking the various pieces become apparent, but where each piece also exhibits unique characteristics that contribute to an overall understanding of the region.[2]

Although much has been written about the shortcomings of federal reclamation in the initial years, few studies to date have focused intensively on

MAP 1.3. The Colorado River drainage basin.

an individual federal irrigation project in order to shed light on the myriad unanticipated environmental, economic, and social challenges with which the government was confronted in the process of bringing arid lands under irrigation and settlement.[3] Yuma amplifies the often conflicting nature of several federal policies, including those governing public land disposal, water rights, and the allotment of Indian reservations, all of which became entangled in the web of early federal reclamation efforts. Yuma therefore embodies both attributes of the final frontier—irrigable and Indian land. Furthermore, a detailed examination of the Yuma Project provides background for an informed understanding of some of the issues involving the Colorado River that would result in unforeseen troubles decades later, including those involving Mexico, Metropolitan Southern California, the Imperial Valley, and the Salton Sea.

Federal Reclamation and Indian Allotment

One objective of this study is to examine the relationship between federal reclamation and the allotment policy, a topic largely overlooked by geographers. In order to make the Yuma Project an economically viable enterprise, the federal government justified the expropriation of the irrigable portion of the Fort Yuma Indian Reservation by employing the Dawes General Allotment Act, a measure approved by Congress in 1887 aimed at "civilizing" Native Americans. The allotment policy was designed to bring Indians into the American mainstream by turning them into successful farmers through the process of allotting them individual plots of ground carved out of their reservation, and opening "surplus" reservation lands to non-Indian farmers who would then serve as positive role models. But the seemingly well-intentioned Allotment Act became an expedient mechanism whereby large chunks of the final frontier—irrigable Indian reservation land held in trust by the federal government—could legally be transferred to non-Indian homesteaders. By making Indian land far more attractive to whites, federal involvement in irrigation not only played a large part in the campaign to amalgamate Indians into white society but also offered new opportunities to exploit Indian resources.[4]

Pisani, for example, notes that in the context of reclamation, "irrigation changed from being the handmaiden of the campaign to 'civilize' the Indians into one of its greatest enemies in the period from 1891 to 1920." He discusses how Indian land and water rights became captive to experiments in federal reclamation by providing the capital to get many government irrigation projects started. That is, Native Americans often subsidized reclamation by giving up reservation land while, in most cases, receiving little in return. Paradoxically, it was the Department of the Interior, through its Reclamation Service, that encouraged the expropriation of Indian land and water resources in the West, while simultaneously, through its Office of Indian Affairs, it was responsible for the protection of those resources. Pisani concludes that it was the enormous political influence within the executive branch of the Reclamation Service, an agency whose primary objective was to provide as much land to white farmers as it could, and one that had little interest in the welfare of Native Americans, that ultimately forced the Indian Office and the Justice

Department to relax their trusteeship over Indian natural resources. Daniel McCool, in his book *Command of the Waters,* also supports the notion that political expediency, if nothing else, dictated that the Reclamation Service favor the dominant white majority, the constituency for which the service was originally established to serve.[5]

Ironically, the first agency to construct irrigation projects on federally managed land was the Office of Indian Affairs, not the Reclamation Service. These projects originally involved an attempt to keep agricultural Indians on their land and persuade nomadic Indians to settle in one place.[6] As early as 1867, Congress made its first federal appropriation for Indian irrigation in order to encourage the concentration of Indians along the lower Colorado River on the newly established Colorado River Reservation. This project, however, was carried out with little forethought and was abandoned after a few years.[7] It was not until the early 1890s, after the adoption of the Dawes General Allotment Act, that regular congressional appropriations for Indian irrigation projects began to be approved.[8] The Indian Office then initiated several small projects, but in nearly every case the work proved unsatisfactory.

Hence, by the time the Reclamation Act became law in 1902, the Indian water development program had made little progress, and a struggle therefore ensued between the Indian Office and the Reclamation Service for control over Indian land and water. The conflict was "fed by the reluctance of the supremely confident officers of the Service to share their mission with any other agency," and in 1906 the Reclamation Service proposed taking over the largest Indian water projects.[9] The following year an interagency agreement was adopted allowing the service to assume direct responsibility for Indian irrigation projects. The commissioners of Indian affairs who served under Roosevelt knew full well that reclaiming the arid West took precedence over Indian affairs with both the secretary of the interior and the president; they therefore failed to uphold their mandate to protect Indian resources by acquiescing to the prevailing political mood within the Roosevelt administration. Moreover, it was assumed that the Reclamation Service could do a better job building and maintaining large-scale hydraulic works than the Indian Office, while a coordinated irrigation policy also would prevent duplication of effort and promote the most efficient use of natural resources. Because of the scarcity of water and arable land in the arid West, irrigation was subsequently extended to many reservations, including the Pyramid Lake Reservation in Nevada, the Gila River Pima Reservation in Arizona, the Yakima Reservation in Washington, the Crow Reservation in Montana, the Uinta Reservation in Utah, and others.[10] These irrigation projects were carried out primarily because of threats to Indian water supplies as well as white demands for surplus reservation land.[11] The case of the Quechan (Yuma) Indians, however, differs somewhat from most reservation Indians in that the latter were involved in irrigation projects sanctioned by and financed through the Office of Indian Affairs, while the Quechans were the first Indian group to play an integral role in the development of a Reclamation Service project specifically intended for non-Indian settlement. To what extent did the Quechan experience conform to or differ from that of other Indian groups?

Robert L. Bee, in *Crosscurrents Along the Colorado*, discusses the impact of a century of federal Indian policy, including the Allotment Act, on the Quechans' economic and social well-being. His study is set within a framework invoking the notion of internal colonialism, a process that "results in a political and economic subordination and exploitation of the colonized by the colonizers." Under such circumstances, Bee notes, there always exists a "profound discrepancy . . . between the declared humanism of national policy and the actual impact of that policy at the local level."[12] With respect to Quechan allotment, this narrative does not refute Bee's thesis, but his primary reliance on information collected from informants on the Fort Yuma Indian Reservation over a period of years tends to tell the story from the Indians' point of view only, causing his coverage to be rather unbalanced. His failure to examine Reclamation Service documents in the context of allotment leaves much of the story untold.

Emily Greenwald analyzes allotment from a somewhat different perspective. Employing geographic methodology, she argues that the Allotment Act had an implicit spatial agenda, that it functioned as a tool of spatial control, designed to alter Indians' spatial relations in fundamental ways. She maintains that a central goal of assimilation policies was to make Indians spatially like Euro-Americans, whereby Indians who occupied land as collective families, bands, or tribes would relinquish this lifestyle for individually owned property. Allotment failed, according to Greenwald, because Indians selected their allotments with agendas other than assimilation in mind. For example, the Nez Perces and Jicarilla Apaches, the two tribes that she examines to support her thesis, resisted assimilation by using allotment selection to assert their own ways of ordering their community. Unlike the Nez Perces and Jicarilla Apaches, however, the Quechans were included in a

Reclamation Service project from its inception, and they therefore had little control over the selection of their individual allotments. Nevertheless, the Quechan experience supports Greenwald's conclusion that "federal policies did not create a uniformity of Indian experience that permits us to make easy generalizations."[13]

David Rich Lewis has studied the Northern Utes, the Hupas, and the Tohono O'odhams, three tribal groups who occupied reservations with different environmental characteristics in the subhumid West. He analyzes how cultural structures and environment influenced each group's response to the same directed change, namely, the Allotment Act. He finds that each group experienced and weighed the federal government's agrarian civilization program against its own cultural and environmental background, and that each responded in culturally consistent ways to shape the nature of that change.[14] Based on his three case studies, Lewis concludes that most Indian groups responded to allotment by selectively adopting aspects of white society and economy as necessary and desirable while resisting other elements requiring more radical changes. As we shall see, the Quechans accepted the alien concept of leasing their allotments to non-Indians largely because the expectation of their becoming highly capitalized irrigation farmers was too radical a change from their traditional overflow irrigation practices for them to embrace. In spite of the injurious effects of allotment, by selectively resisting certain elements of white society, the Quechans, as with most other impacted Indian groups, have been able to maintain their identity as a separate people.

From Indigenous to Industrial Irrigation

As will be seen in more detail in chapter 2, the Quechans originally practiced a primitive form of irrigation known as flood recession farming along the lower Colorado River. The Quechans and other nomadic bands of Native Americans in the Southwest gradually acquired an understanding of the relationship of plant growth to moisture, temperature, soil fertility, and seasonal changes that eventually led them to settle in permanent villages near stream courses. They practiced a natural form of irrigation that obviated the need to develop complex water-control schemes by utilizing the summer seasonal spillover on the swollen Colorado River to inundate an otherwise dry floodplain. The residual soil moisture, which remained high for a prolonged period after the flood receded, allowed crops to grow and mature during the rainless growing season. Furthermore, the fertility of the floodplain was enhanced by the annual overflow as a new layer of silt containing beneficial nutrients was deposited each year.[15] Indigenous irrigation practices such as flood recession farming did not require great labor or technological inputs, and, unlike more technologically advanced forms of irrigation, were ecologically and socially sustainable over the long term. Indeed, most of the farming practices employed by aboriginal North Americans were sustainable over the long term.[16]

An exception involved the Hohokam, who inhabited the Gila River and Salt River valleys in southern Arizona. Flood recession farming no doubt eventually led to the elaborate artificial irrigation works built by the Hohokam, who created an extensive canal system that irrigated thousands of acres and supported a complex culture spread across a network of permanent villages. Floodplain settlement began as early as 300 B.C., and by A.D. 1200 ten distinct canal systems, comprising at least 150 miles of large canals, had been constructed in the vicinity of present-day Phoenix.[17] The intensive artificial irrigation employed by the Hohokam allowed them to expand the area of crop production and to increase yields on existing croplands in the Salt and Gila river valleys, which in turn led to higher population densities. Around A.D. 1400, however, this complex of irrigation works was abandoned, possibly because irrigation over the centuries induced water logging of the soil, thereby drowning the roots of crops and causing alkali problems, rendering much of the land unfit for cultivation. Hence, along with the intensification of irrigation came salinization, which poisoned the farmers' fields. Worster remarks that the Hohokam suffered from the political and environmental consequences of bigness.[18]

Flood recession farming, particularly along larger rivers such as the Colorado, is no longer common in the American West.[19] Instead of respecting the seasonal rhythm of rivers by developing symbiotic relationships with their rise and fall, we have invested in massive industrial irrigation systems in our attempt to achieve control over them. Transitional between indigenous and industrial modes of irrigation was that created by pioneer families and mutual irrigation companies working to build small-scale irrigation works across the arid West. "Through plowing, planting and irrigating, [family] farmers . . . tried to impose their designs on nature" in order to liberate themselves from their dependence on precipitation. However, "Instead of shaping the land to their liking, the irrigators often found themselves reacting to natural processes that they could not master and that they dimly understood."[20] As we shall see, with the onset of industrial irrigation, the problems associated with a lack of understanding of natural processes in an arid environment were magnified manyfold.

Industrial irrigation has inevitably led to the industrialization of agriculture, relying on mechanization, increased energy and capital inputs, large-scale physical infrastructures, centralized management, and crops bred to be more responsive to agrochemicals and increased water supplies. Worster points out that wherever intensive, large-scale irrigation has appeared, farming has quickly become a factory operation, "mass-producing for a mass-consuming market."[21] Although this type of irrigation farming is expensive, the total costs of construction and water delivery of large-scale irrigation projects are often hidden because they are subsidized by government tax revenues. Few farmers could afford it if there were no subsidy. Industrial irrigated agriculture, therefore, appears more economically successful than it would be if the benefiting farmers had to assume complete costs at true market values.[22]

In the United States, we have called on our technological expertise to dam and divert virtually every major western river in order to nourish thirsty fields "dependent for their productivity on large machines, fossil fuels, chemical fertilizers, insecticides, and herbicides." According to Worster, this mode of water manipulation works to achieve a control over nature that is "unprecedentedly thorough." Consequently, we are caught in an irrigation conundrum whereby indigenous systems of irrigation such as flood recession farming most likely could have sustained relatively small populations indefinitely, whereas industrial irrigation is able to support much larger populations but is able to do so only because of "its willingness to destroy itself."[23] In other words, we have elected to substitute long-term sustainability for maximum productivity in the short term.

The territory encompassed by the Yuma Project is representative in this regard, for it illustrates not only how "there once were men capable of inhabiting a river without disrupting the harmony of its life" but also how humans, through the control of water in the modern American West, have attempted to achieve the technological domination of nature.[24] In the chapters that follow, the story of how one corner of the southwestern United States was transformed from an indigenous mode of natural irrigation to an industrial mode of irrigation based on megascale water-control technologies will unfold. But before I delve into detailed developments at Yuma, a discussion of the rationale underlying the approval of the Newlands Reclamation Act, which represents the onset of America's industrial irrigation program on a massive scale, would prove useful.

Prelude to Federal Reclamation

Early in the twentieth century, when the national reclamation law was approved by Congress, an estimated 7.5 million acres were already being irrigated in the West, most of which resulted from the hard work and cooperative efforts of pioneers with limited means.[25] By this time, however, nearly all the land that was available for irrigation utilizing small-scale irrigation works resulting from the toil of hardy pioneers was occupied; in order to bring additional lands under irrigation, more elaborate and expensive storage and diversion works would be required. The overwhelming majority of corporate irrigation schemes that attempted to provide such irrigation works had been failures, as many were planned, engineered, and promoted by incompetent or speculatively inclined persons having insufficient capital to build the large dams, storage reservoirs, and high-line canals necessary to irrigate large tracts of land.

Congressional legislation specifically designed to encourage farming in the arid West began with approval of the Desert Land Act in 1877. This land measure, which supported the traditional position that land development should be left to private initiative, allowed for the purchase of 640 acres at a cost of $1.25 per acre, with the provision that the land would be irrigated within three years. Twenty-five cents per acre was paid at the time of entry, while the additional dollar per acre was due on proof of reclamation. Congress believed that relatively large full-section parcels were necessary in order to make investment by individuals in irrigation works economically feasible. Although designed to aid homesteaders, the Desert Land Act called for no residency requirement and was therefore generally unsuccessful in placing bona fide settlers on the land. Instead, it was widely abused by nonresident entrymen, such as eastern investors speculating on increased land values in connection with proximity to water, and by stock raisers who used the land measure as a holding tactic in order to control grazing lands bordering western streams.[26]

In 1891, the Desert Land Act was revised, reducing the maximum parcel size to 320 acres while requiring entrymen to be residents of the state in which they were filing and to submit plans for reclaiming their tracts. The law also authorized the secretary of the interior to grant canal rights-of-way across the public domain to irrigation companies. But most private irrigation projects were partial or complete failures. In many instances, irrigation projects were begun without sufficient capital and work was suspended before water was delivered. Consequently, more than 90 percent of private canal projects were in or near bankruptcy by 1900.[27]

The necessity for congressional action on a more comprehensive scale was initially set forth in 1878 in John Wesley Powell's *Report on the Lands of the Arid Region of the United States*. A decade later a more intensive investigation of the extent to which the arid region could be redeemed by irrigation was included in the congressionally authorized Powell Irrigation Survey.[28] Powell, then head of the U.S. Geological Survey, was instructed to survey potential reservoir sites and irrigable lands throughout the West. Land speculators, however, followed the survey parties into the field, filing on tracts before they could be reserved, which led the General Land Office to close the public domain until Powell finished his work. Westerners vehemently protested this action, and, in 1890, Congress restricted withdrawals to actual reservoir sites and cut the irrigation survey's funds. Hence, Powell's plan to carefully inventory the West's water resources and to control development was rejected in favor of identifying water storage sites, which in turn was accompanied by unrestricted alienation of adjacent lands.[29]

Soon after the initiation of the Powell Irrigation Survey, there was sufficient congressional interest in irrigation to establish a Select Committee on Irrigation and Arid Lands in each house, and the U.S. Senate quickly authorized its committee of seven senators to look into the best way of reclaiming the country's arid lands. Within a span of just one and a half months, the select committee held meetings and collected several volumes of testimony in nearly fifty cities scattered across the West.[30] Popular interest in irrigation stimulated by the Powell Irrigation Survey and the *Report of the Senate Committee on Irrigation* quickly led to a series of National Irrigation Congresses. William Ellsworth Smythe, a Nebraska journalist who had attracted public attention with a series of articles on irrigation in the *Omaha Bee,* was made chairman of a committee to help organize the first National Irrigation Congress, which met in Salt Lake City in September 1891. Smythe also founded the journal *Irrigation Age* and authored *The Conquest of Arid America,* the bible of the reclamation movement. According to Michael Robinson, Smythe "was the West's first spokesman to develop a common focus and ideology for the irrigation movement."[31]

Subsequent congresses were scheduled annually at various western cities. At its ninth session, held in 1900, it was recommended that the work of building reservoirs on western rivers be done directly by the federal government. It should be noted, however, that Powell, who inspired interest in irrigation and western land settlement with his report on the arid region of the United States and whose irrigation survey stimulated interest in a more comprehensive program to develop the West's irrigation potential, never advocated government construction or operation of reclamation works. Instead, he envisioned farmers and ranchers organizing themselves in self-governing irrigation and pasturage districts, under congressional legislation, to manage settlement, water development, and the use of natural resources.[32] Powell believed that the sole responsibility of government should be to furnish the districts with the best scientific information it could.[33]

Prior to approval of the Reclamation Act, an attempt had been made by the federal government to interest arid land states in their own land reclamation projects. Through passage of the Carey Act in 1894, a measure introduced by Wyoming senator Joseph M. Carey, one million acres of desert land in the public domain would be donated to each of the ten arid land states if they would devise some means to reclaim their grant. The states were to contract with private construction firms for building the irrigation works that would be reimbursed through the sale of water rights. The land grants would then be disposed of to farmers in tracts no larger than 160 acres, with the understanding that at least 20 acres of each quarter section would be cultivated within ten years. The law, however, was a dismal failure. Five years after the Carey Act's approval, only Wyoming—the home state of the senator who sponsored the bill—had requested a grant under the act, and ten years after its passage only 11,321 acres had been patented under its provisions.[34]

The disappointing failure of the Carey Act to encourage western states to establish viable reclamation programs convinced Congress that it was time for the federal government to intercede, that is, to begin to make beneficial use of the volumes of water that were "running to waste" in the West.[35] The immense expense of river control and irrigation development and the interstate character of most major streams, as well as the federal government's position as the largest landowner in the West, all argued for the need for federal involvement.[36] Hence, on June 17, 1902, Congress gave its approval to a reclamation bill sponsored by Representative Francis G. Newlands of Nevada, launching "potentially the biggest public works program in American history." Its passage represented the culmination of efforts to adapt national land and water policies to western conditions. Worster calls the Reclamation Act "the most important piece of legislation in the history of the West."[37]

The Newlands Reclamation Act

When the Newlands Reclamation Act was passed, a vast body of information concerning the irrigation possibilities of the West had already been gathered in connection with Powell's irrigation survey, and by the hydrographic branch of the U.S. Geological Survey. General studies of streams, watersheds, irrigable lands, and potential dam and reservoir sites had been completed for much of the West. The Reclamation Service supplemented this information with more intensive surveys of the most promising irrigation projects, and began to prepare plans and specifications for construction. Project recommendations were then forwarded to the secretary of the interior. Initially, the Reclamation Service, which would oversee federal irrigation, was made a division within the hydrographic branch of the U.S. Geological Survey, and Frederick H. Newell, a civil engineer who had served as head of the hydrographic branch since 1890, was named chief engineer of the service. In 1907, the Reclamation Service was separated from the Geological Survey and made directly responsible to the secretary of the interior. Newell at that time became director of the Reclamation Service, a position he retained until 1914.

On March 14, 1903, Ethan A. Hitchcock, Roosevelt's secretary of the interior, approved the construction of five Reclamation projects, including the Truckee-Carson (renamed Newlands) Project in Nevada, the Salt River Project in Arizona, the Gunnison (renamed Uncompahgre Valley) Project in Colorado, the Sweetwater (renamed North Platte) Project in Nebraska, and the Milk River Project in Montana, while another six projects were approved in 1904, nine in 1905, one in 1906, and one in 1907. Hence, a total of twenty-two projects were authorized within a few years after passage of the reclamation law, twenty-one of which were carried to completion (map 1.4).[38]

Prior to the Reclamation Act's approval, President Roosevelt had cautioned that "it would be unwise to begin by doing too much, for a great deal will doubtless be learned, both as to what can and what can not be safely attempted, by early efforts, which must of necessity be partly experimental in character."[39] But within two weeks after passage of the Reclamation Act, Roosevelt had reconsidered and instructed Interior Secretary Hitchcock to spread the work among the western states as widely as possible, instead of embarking on a few experimental projects. Political expediency governed Roosevelt's decision to push construction of as many projects as possible, as

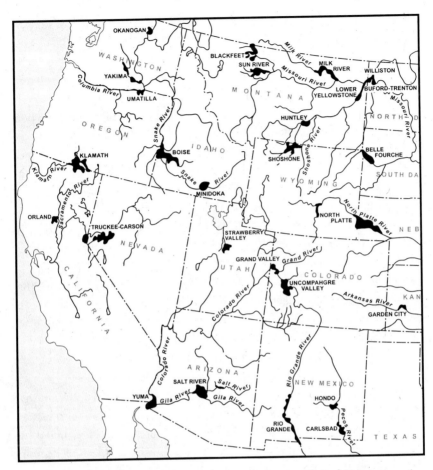

MAP 1.4. Early federal irrigation projects in the western United States, 1910. A few projects shown on the map were never initiated. (Based on F. H. Newell, "Progress in Reclamation of Arid Lands in the Western United States," 182)

rapidly as possible.[40] Since Section 9 of the Reclamation Act stipulated that proceeds derived from the sale of public lands would be spent for reclamation in the public land states in proportion to the revenues derived within each, every western state believed that it was entitled to at least one project. Therefore, both the president and his interior secretary, who was in charge of project selection, were pressured to build a reclamation project in each of the public land states.

Since the federal reclamation program was founded on the belief that storing floodwaters and building canals would foster western settlement, its primary goal was to use the West's surplus water to carve new farms out of the public domain.[41] When the Reclamation Act was approved by Congress, President Roosevelt had emphasized that it was not enough for the government merely to regulate the flow of western streams; instead, the primary objective was to dispose of irrigable land in the West to settlers who would build homes on it.[42] But Interior Secretary Hitchcock's eager authorization of projects quickly placed a severe financial strain on the Reclamation fund. In fact, in less than a decade after passage of the Reclamation Act, the Reclamation Service was in serious financial trouble, leaving little money available to assist settlers in the formative years when they would have to carry out the difficult and costly work of clearing and leveling the land, digging irrigation ditches, building roads and houses, and transporting crops to distant markets.[43] Reisner views the early Reclamation program as much a disaster as its dams were engineering marvels.[44]

Under the new program the federal government would construct the necessary large-scale irrigation works, while the public lands included in the Reclamation projects would be disposed of free under the Homestead Act. Settlers, however, were required to reimburse the government for construction of the irrigation works, but these charges would be spread over a ten-year period, interest free. In the meantime, the new engineering works would be financed out of proceeds derived from the sale of public lands in the western states. Those homesteading on surplus Indian lands also had to reimburse the Indians for their irrigable parcels at a price determined by the federal government. After construction was paid off, the Reclamation Act called for the management and operation of the irrigation works to pass to the owners of the irrigated lands located within the federal projects. It all seemed fairly simple from the standpoint of the Reclamation Service— construct the dams and levees and excavate the canals that would bring water to the remaining irrigable lands in the West, and vast desert tracts would eventually become agricultural oases resulting from the hard work of rugged pioneer homesteaders.

A Look Ahead

As the events unfold at Yuma it will be seen that nothing could have been further from the truth, for Yuma clearly demonstrates the experimental nature of the federal government's early attempts to reclaim irrigable lands in the arid West. In addition to the complex issues surrounding the Fort Yuma Indian Reservation, there were other problems—many unforeseen—that

FIG. 1.1. Overflow lands near Yuma prior to the onset of the Yuma Reclamation Project, ca. 1902. (Reproduced from J. Garnett Holmes et al., "Soil Survey of the Yuma Area, Arizona," plate 54)

FIG. 1.2. Yuma Reclamation Project lands today showing lettuce and date production. (Photo by the author)

had to be dealt with at Yuma. For example, one of the fundamental difficulties associated with federal reclamation involved how to handle irrigators who had already taken advantage of various prior federal land measures designed to encourage settlement of the public domain. In fact, two-thirds of all the land on the early projects was already in private ownership when they were authorized.[45] In this regard, the Yuma Project was no exception, for the Yuma Valley, which constituted approximately two-thirds of the original Yuma Project lands, was largely in private hands before the project was approved. Decisions also had to be made concerning what size limitations should be placed on landholdings that had been settled prior to approval of the Yuma Project, how the allotment process should be carried out, what should be the appropriate size of the farm unit for non-Indians on the surplus irrigable lands of the reservation, what types of crops should be grown and where they should be marketed, how negative environmental impacts associated with irrigation, such as salinization and drainage, should be dealt with, and so on.

The Yuma Project, which encompassed land once covered with mesquite and wild grasses, and successfully farmed by indigenous peoples, has grown into one of the most productive farming regions in the Southwest (figs. 1.1 and 1.2). But this transformation has been achieved only through a long period of trial and error, which provoked serious environmental repercussions, and social and economic distress. My intent is to examine the untold difficulties associated with federal reclamation at Yuma, and the manner in which they were overcome, in order to better understand the processes associated with the government's effort to expand the bounds of irrigation and homesteading on the nation's final frontier. As William Wyckoff notes, "Western historical geographers still have many stories to tell about how Western places have changed through time."[46] This study tells the story of how one small but significant part of the nation's final frontier became the place that it is today.

2

Quechan Land

According to the Quechan creation myth, Kumastamxo, the younger of two great spiritual leaders of the Yuman tribes, lived with his various peoples atop a flat-topped mountain known as Avikwamé, the place of creation. The several Yuman tribes then all descended from the top of Avikwamé to inhabit their respective territories. The Quechan, however, followed *xam kwatcán,* meaning "another going down," or a trail different from the others. Hence, it is assumed that the Quechans adopted their tribal name from the word *kwatcán.*[1]

Outside of their tribal setting, the Quechans have more commonly been known as the Yumas. *Yuman,* a branch of the Hokam family of languages, is the linguistic grouping to which the Quechans belong. It encompasses the Indian tribes occupying the bottomlands and delta of the lower Colorado River from the Mohave tribe, near present-day Needles in the North, to the Cocopa, who in aboriginal times inhabited the area near the Gulf of California (map 2.1). In addition to the Mohave, Quechan, and Cocopa, several other tribes of Yuman lineage once occupied the banks of the lower Colorado, including the Halyikwamai, Kohuana and Halchidhoma, while the Maricopa inhabited the lower Gila Valley. Presumably, it was the non-Yuman Pima, Papago (Tohono O'odham), and other tribes who referred to the Quechans as *Yum,* but C. Daryll Forde believes it is possible that the Spaniards, who were using the term *Yuma* in the eighteenth century, introduced it among the Indians.[2] Regardless, Quechan is the proper name of the river tribe that has traditionally inhabited the lower Colorado River region near the Gila River confluence.

River Civilizations

The tribal homelands of the Cocopa, the Quechan, and the Mohave had the advantages of being larger and more productive than those occupied by the other tribes. These three tribal nuclei, being sufficiently distant from one another, permitted the rise of three strong and independent river

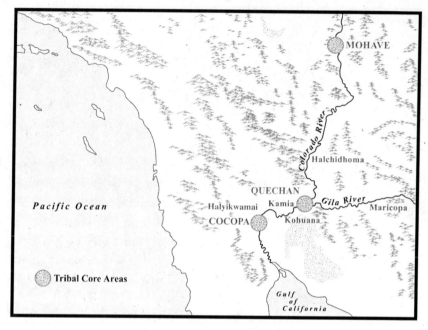

MAP 2.1. Yuman river tribes of the lower Colorado River. (Based on A. L. Kroeber, *Cultural and Natural Areas of Native North America,* Map 1A: Native Tribes of North America; and Raymond W. Stanley, "Political Geography of the Yuma Border District," 1:51)

tribes, who expanded at the expense of the smaller river tribes found in the area.[3] The centrally located territory of the Quechan tribe was the pivotal point in intertribal strife, for it was at the Colorado-Gila confluence that dispersal of the river tribes apparently emanated. In the nineteenth century, the Halchidhoma, Halyikwamai, and Kohuana were expelled from the Colorado region.[4] The displaced tribes migrated eastward along the lower Gila Valley to join the Maricopa Indians. In contrast to the exodus of these tribes, a small group of eastern Diegueños, the Kamia, moved into Quechan territory prior to 1850, and eventually came to locate in the area south of Pilot Knob. The Kamia recognized the territorial claim of the Quechans to the land they occupied and lived there as guests of the Quechan. They also served as a buffer group between the Quechans and the Cocopa. Traditional enmities between individual tribes were no doubt related to competition over the same general type of habitat, for the Yuman river people recognized themselves to be bottomland agriculturalists and considered the floodplains their exclusive domain.[5] But traditional alliances among tribes also existed, as the powerful Quechan and the Mohave river tribes and several smaller groups typically allied themselves against the Cocopa and its tribal allies.

At one time, the territory held by the Quechans might have extended along both banks of the Colorado River, from several miles south of the Gila confluence to approximately sixty miles north, near present-day Blythe, but its undisputed tribal core area focused on the strategic crossroads at the confluence of the Colorado and Gila rivers (map 2.1).[6] From this center the tribal territory expanded outward in concert with population pressure within the core area and conversely with the resistance of neighboring tribes.[7] The Quechan, like other Yuman tribes, regarded themselves as national entities with considerable tribal solidarity, and their land was thought of as a country. Fred Kniffen notes, however, that it is difficult to determine precisely the areal divisions of the several tribes because of shifting locations due to intertribal warfare.[8] Yet shifting groups did not mean shifting cultures, as unity of language and culture characterized all the Yuman tribes.

Malcolm Rogers postulates that about twelve hundred years ago the ancestors of the Yuman people drifted eastward across the desert from Southern California to inhabit the lower Colorado River valley. Although originally seed gatherers, the groups that settled near the Colorado-Gila confluence turned to agriculture and pottery making; they therefore retained a California type of culture, but adopted "new implements to meet the exigencies of harvesting and preparing the foods of a different ecologic zone."[9] Hence, one of the common threads woven among the Quechan and other river tribes is the close relationship that developed between them and the presence of the Colorado River. In addition to their tribal name, the origin and significance of the Colorado are revealed in the Quechan creation myth. It is told that Kumastamxo plunged a wooden spear into the ground to cause the river to gush forth. Wherever he held the spear flat, the river channel became broad; where he held it sideways, the river channel was narrow. Near present-day Yuma he cut down the mountains to let the river through, and when the weather grew hot, it would rise and overflow its banks. The Quechan and other river civilizations depended on the Colorado's annual overflow to develop their specialized culture. According to A. L. Kroeber, it was a culture that entailed "consequential agriculture depending wholly on river bottom-land flooding, not at all on rains or artificial irrigation."[10]

The American Nile

The Colorado River, after emerging from its headwaters and tumbling down the west-facing slopes of the Rocky Mountains, etches itself into the Colorado Plateau before crossing the southern Basin and Range, one of the hottest and driest expanses found anywhere in North America (see map 1.3). Seventeen hundred miles distant from its headwaters, after draining portions of seven states and a small part of northwestern Mexico, the Colorado River empties into the Gulf of California south of Yuma. Along its lower course the Colorado marks the boundary between parts of Nevada and Arizona and California and Arizona, and, for a few miles before crossing entirely into Mexico, it forms the border between the southwest and northeast corners of Arizona and Baja California, respectively. The last canyon through which the Colorado River has cut is located about ten miles above Yuma where it separates the Chocolate and the Laguna mountains. During the late Pleistocene, the river entered the Gulf of California at this locality and began building its delta.[11] From this point, the river today meanders for a distance of about one hundred miles across its delta, south from Yuma, before reaching the gulf (map 2.2).

In the vicinity of Yuma, the Colorado floodplain is bordered by terrace or mesa edges that rise abruptly fifty to one hundred feet above the floodplain floor. A striking feature found on the mesa extending to the west of Yuma is a belt of large, active sand dunes, formerly known as the sand hills (today called the Imperial Dunes). Averaging three miles wide and trending southeastward without a break for a distance of thirty-five miles, the sand hills spill across the international boundary for a short distance into Mexico (see map 2.2). Until well into the twentieth century irrigation canals, railways, and roads had to swing far to the north, or to the south into Mexico, to avoid this barrier. The Colorado River also was forced to meander around this obstruction during times that it flowed north into the Salton Sink.

With its delta in the desert, the Colorado has no parallel in the Western Hemisphere, but in the Old World the lower Nile Valley is a fitting correlative. The Nile, like the Colorado, rises in a distant, mountainous country, while its lower course traverses a subtropical and nearly rainless desert. For long distances along their lower courses, both rivers have deposited a narrow ribbon of fertile soil in the midst of the barren deserts through which they flow, and both rivers are subject to an annual summer rise coming at

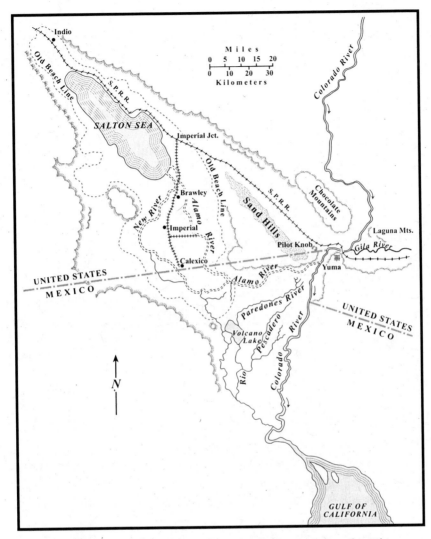

MAP 2.2. The Colorado River delta region. (Based on C. E. Grunsky, "The Lower Colorado River and the Salton Basin," 15)

an opportune time for irrigation.[12] It is because of these similarities that the Colorado River is often referred to as the American Nile.

Even though the Colorado is situated in a severe climatic desert, Kniffen has pointed out that because of the presence of the river, it is not a desert in a vegetative sense.[13] The contrast between the dry mesas flanking the luxuriant floodplain in aboriginal times was striking. A vigorous

hydrophytic and mesophytic plant growth including dense groves of cottonwood, black willow, and mesquite, and impenetrable thickets of arrowweed
six to eight feet tall were found on alluvial soils subject to overflow, whereas
a more open xerophytic landscape composed of creosote bushes, ocotillo,
palo verde, ironwood trees, and some cacti characterized the poorly watered
sandy and gravelly mesas. The cottonwoods and black willows were found
near the water's edge, rushes, tules, and canes occupied the sloughs, while
arrowweed formerly dominated the greater part of the overflow land. The
less water-tolerant mesquite grew most abundantly along the outer margins
of the floodplain beneath the mesa bluffs on land that was rarely flooded.

Flood Recession Farming

The phrase "consequential agriculture" coined by Kroeber actually involved
flood recession farming, or the Quechans' use of the annual floodwaters of
the Colorado River for farming, since the region's average annual rainfall
of less than three and a half inches was inadequate for successful crop cultivation. The annual flood inundated the overflow areas, arriving generally
in mid-May and subsiding by midsummer. The highest waters normally occurred between mid- and late June. Important from the standpoint of Indian
agriculture was the moisture that seeped into the soil and the replenishment
of soil fertility as a result of the overflow. The volume of silt carried by the
Colorado River is enormous, and the continuous use of the floodplain lands
without exhausting their fertility was made possible because of the fertilizing elements deposited by the annual overflow.

Each year in the spring prior to the overflow, small plots subject to flooding were selected for cultivation and cleared with axes and knives. The
ground was tasted, and if the presence of salt was detected, it was rejected.[14]
Preferred areas for planting were those over which the floodwaters moved
rapidly enough to deposit coarse sediment, for areas flooded by still water
tended to be characterized by heavy soils subject to cracking. After the river
subsided, holes were dug in the moist earth with a digging stick to a depth
of several inches. If the ground was easily penetrated, it was an indication
of a good place to plant. Where the land was suitable, fast-maturing seeds
of the standard aboriginally cultivated crops—maize, beans, and squash—
were thrown into the holes and covered with damp earth. Melon, gourd,
and pumpkin seeds were also planted. Since crops had to germinate and
grow to maturity almost entirely on moisture furnished by the river during
its period of flood, seeds were planted separately and in the order of their

FIG. 2.1. Home of a Quechan family, date unknown. (U.S. Bureau of Reclamation photo, courtesy of the Yuma Area Office)

soil-moisture needs and the lengths of their maturation periods. Wheat and barley were early introduced or diffused into the area from Spanish sources, and although they were not preferred as foods, their short growing season permitted them to mature following off-season flooding, thus enhancing the security of food supplies.[15] The Quechans followed the receding water, planting higher ground first and lower lands later in the freshly exposed mud immediately after the flood subsided. Nothing further was done by the Quechans except for the removal of weeds, either by hoeing or by pulling weeds by hand, until the products ripened and were harvested beginning in September and continuing into November. Yields generally were dependable and abundant. A typical family cultivated one or two acres, producing far more food than their needs.[16]

The Quechans had at their disposal at least thirty thousand acres that were sometimes subject to flood, but probably no more than one thousand acres were ever cultivated.[17] Their horticultural products were supplemented with wild seeds, fish, and some game, although fishing and hunting were secondary to agriculture and gathering. Men hunted and fished, while wild plant foods were gathered by women, the most important being mesquite pods. Mesquite pods and beans were an especially important element of the Quechan diet, which they ground and mixed with water until the consistency of mush, or cooked over heated stones to make a kind of bread.[18] The sources of Indian subsistence in terms of gross caloric intake have been estimated to be 40 percent from cultivated crops, 40 percent from wild plants, 10 percent from semicultivated grasses and herbs, 7 percent from fish, and 3 percent from game.[19] The most serious difficulties arose in years when the river failed to rise and temporarily inundate prospective planting grounds, or when a late second flood washed out half-grown plantings. At such times severe scarcity of food would result, and sustenance had to be provided largely from wild plant foods.

The reliance on flood recession farming based on the river's overflow supplemented by the gathering of wild seeds supported relatively high population densities among the river tribes along the lower Colorado. The Quechans, who probably numbered more than four thousand prior to European contact, lived in scattered settlements, or *rancherias,* small, loose clusters of perhaps a half-dozen to a dozen huts occupied by friends and relatives, near their patches of farmland.[20] Family dwellings generally were temporary, as their locations were periodically changed. However, although the distribution of individual dwellings was temporary, the general sites of the major *rancherias* were relatively permanent. Summer houses, which consisted of simple brush shelters, or somewhat more substantial *ramadas,* were located above the overflow land on river terraces (fig. 2.1). The ramadas were constructed by sinking four corner posts into the ground, over which a simple flat roof of arrowweed stalks was placed that provided protection from the burning summer sun where temperatures on occasion could reach 120 degrees in the shade. The winter dwelling sites were in the river bottoms, where a semisubterranean, or pit, house was often used for greater warmth. A pit was dug around which logs were upended. The logs were then covered with poles and brush and plastered over with dirt.[21]

Non-Indian Intrusions

Excepting the consequences of persistent intertribal conflicts, the lives of the Quechans remained largely uninterrupted until the arrival of Europeans. In 1539, Francisco de Ulloa reached the mouth of the Colorado River while exploring the Gulf of California, where he performed a symbolic act of taking possession of the region in the name of the king of Spain.[22] Ulloa did not enter the river himself, and made no contact with the Indians. Although initial contact between the Indians and the Spaniards occurred the following year, when Hernando de Alarcón ascended the river as far as the Pilot Knob area, it was the use of the Yuma crossing at the Gila confluence by the Spaniards, who were traveling from Sonora to their incipient coastal California settlements more than two and one-quarter centuries later, that most directly impacted the Quechans' way of life. Meanwhile, European articles, including the horse, wheat, and millet, had already been disseminated among the Indians. Of all the riverine Yuma groups, the Quechans, because of their strategic location at the Colorado-Gila confluence, received the most intense Spanish acculterative influences in terms of horses, new crops, and ideas about European civilization.[23]

Spain wanted to expand its influence in the Southwest in order to forestall possible encroachment by other colonizing powers, but it lacked sufficient Spanish colonists for this work. The crown therefore turned to the church to win control of the Indians by peaceful persuasion. In January 1774, Captain Juan Bautista de Anza, who had been commissioned to find an overland route from Sonora to Alta California, visited the Quechans. Accompanying him was Franciscan missionary Francisco Garcés, who in 1772 had explored alone along the lower Colorado as far west as present-day Mexicali.[24] Fray Garcés observed that an overland crossing to the Pacific Coast of Alta California would be possible, and he conveyed this information to Anza.[25] Anza and Garcés concluded that the best route to California was north along the Rio Altar in Sonora, then northwestward to the settlement of Sonoita, and from there via the Camino del Diablo to the Yuma crossing.[26] The route to Alta California would then continue westward along the Alamo River (map. 2.3).

After their return to Mexico City, Anza urged the opening of this route, while Father Garcés recommended the establishment of a presidio and mission settlement at the Yuma crossing.[27] Such a colony would be ideally situated about midway along the desert route, where it could serve to provision colonists en route to California, aid travelers in crossing the Colorado River, as well as provide protection against the Indians. In February the Quechans were informed that they were now Spanish subjects. However, the Spaniards made no attempts at settlement for several years, and the Quechans initially developed friendly relations with the Europeans.

In 1780 King Carlos of Spain authorized the founding of a mission colony on the lower Colorado River.[28] Officials in Mexico decided that the colony

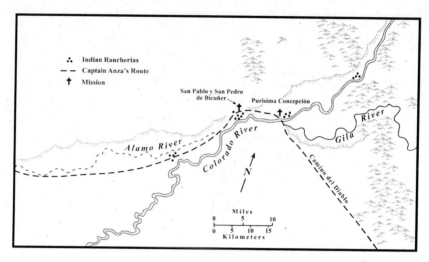

MAP 2.3. Captain Anza's route through Quechan land and location of Indian *rancherias* and Spanish missions, ca. 1780. (Adapted from Raymond W. Stanley, "Political Geography of the Yuma Border District," 1:66)

would consist of two small mission settlements separated by just a few miles. In 1780, mission Purisima Concepción was established at the Yuma crossing, while mission San Pablo y San Pedro de Bicuñer was located eight miles downstream (map. 2.3). Because of the scarcity of available troops, and the friendly contact between the Spaniards and the Quechans, it was decided that a presidio would not be established in the upper delta region to provide protection for the missions. Instead, the two settlements were to be of a mongrel nature, each composed of a few priests, a few lay colonists, and a few soldiers.[29]

But early in 1781, after the two mission settlements had begun to take shape, the Quechans became aggressive. A large party of colonists and soldiers en route to San Francisco thoughtlessly allowed their cattle to graze on fields planted by the Quechans, destroying their crops as well as the mesquite groves that the Indians relied on for food. Meanwhile, the Quechans also had proved unwilling to embrace the new doctrines forced on them by the missionaries. On the morning of July 17, 1781, after most of the armed escort and colonists had departed, the Quechans rose in rebellion to destroy the two mission settlements and expel the Spaniards. In three days of fighting they razed the two missions and killed more than a hundred colonists, including Father Garcés. Only a few women and children survived the attack. The Spaniards suffered no other defeat so crushing and decisive anywhere in the northwestern provinces of New Spain, and, after the Quechan revolt, the Spanish crown determined that the Colorado River posts were not really necessary after all. This event, in conjunction with the Mexican Revolution, caused the Spaniards to abandon the region for good. Although the Quechans were never again to be subdued by the Spaniards, intertribal warfare would weaken them in the following decades, making them less able to meet the threats posed by subsequent intruders.[30]

Beginning around 1827, several parties of fur trappers and explorers reached the Colorado River in Quechan territory, including parties led by James Pattie and George C. Yount. But after the decline of the fur-trapping industry in the mid-1830s, little contact with Anglo-Americans occurred until the latter part of the following decade. Meanwhile, from 1835 to 1845, Mexicans on their way to coastal California were making use of the Yuma crossing, their trespass presumably sanctioned by the Quechans. But unbeknownst to the Quechans, in 1838 a sizable land grant, identified as the Algodones Grant, encompassing most of the Yuma Valley, was made to Fernando Rodriguez by the then renegade "free and independent sovereign state of Sonora."[31] In January of that year Rodriguez presented a petition to the treasurer-general of the state of Sonora stating that he had sufficient means to settle and cultivate a tract of desert land encompassing five square leagues, extending eastward from the confluence of the Colorado and Gila rivers. His petition stated that he would settle and cultivate said lands as soon as the dangers posed by the Indians were subdued. There is no evidence to suggest that Rodriguez ever brought the Algodones Grant under settlement, but permanent settlement of the Yuma Valley was impeded until arguments regarding the authenticity of the grant were heard before the U.S. Court of Private Land Claims, and ultimately the U.S. Supreme Court decades later.

Quechan independence was not threatened until after the Mexican War when in 1848 the United States took over present-day California and the lands north of the Gila River in Arizona, the territory ceded by Mexico by the Treaty of Guadalupe Hidalgo. Subsequent federal surveys of the region, however, made government officials realize that the decision to accept the Gila River as the international boundary was a mistake, particularly in light of the future construction of a railroad line across the Southwest. Hence, the part of Arizona lying south of the Gila River, including the Yuma Valley and the Algodones Grant, was not incorporated into the United States until the

Gadsden Purchase of 1854. In the span of six years, then, between 1848 and 1854, all of Quechan land had been absorbed into the United States. This period also saw an enormous invasion of travelers using the Yuma crossing, but without the consent of the Quechans it was impossible for non-Indians to cross the Colorado River at the Gila confluence. In order "to secure and to hold" this passage, the United States Army sent soldiers to Quechan land.[32] It was during this time that the Quechans were reduced to the status of a conquered people.

The Yuma Crossing

The single most important event to alter the lives of the Quechans was the California gold discovery in 1848, for many of those who flocked to the gold-fields from other parts of the country followed the Gila River to the Yuma crossing to reach their ultimate destination. Thousands more came from Mexico, most of whom trekked along the Santa Cruz River to its confluence with the Gila. In 1849 alone, it is estimated that six to nine thousand Americans and fifteen thousand Mexicans passed through Quechan land.[33] This tremendous influx of transients, with their horses and mules, undoubtedly destroyed much of the pasture and farming land of the Quechans. The Indians, however, attempted to make the best of an otherwise damaging situation by hiring out their services to help the travelers cross the Colorado River. The Quechans, who were excellent swimmers, made rafts by lashing together bundles of reeds that they then pulled by swimming across the quarter-mile-wide river. In this manner they helped ferry the goods of the forty-niners across the river, while others swam the herds of animals to the other side.

According to Jack Forbes, the Quechans believed that since the crossing had always belonged to them, they alone possessed the "franchise" of transporting goods across the river.[34] But the Americans did not recognize the Quechans' ownership of Quechan land, let alone the Yuma crossing, and began to initiate ferry services of their own. This led to hostilities between the two groups that resulted in the death of a white ferry operator. A California militia was subsequently organized to punish the Quechans. In September 1850, the volunteers marched to the Colorado River, destroyed the Quechans' fields, and put an end to their ferry business. After the militia left, the Indians retaliated by taking a ferry operator and his men hostage. Three companies of U.S. troops, commanded by Maj. Samuel P. Heintzelman, came to the rescue of the hostages and established Camp Yuma on the west bank of the Colorado River, one mile west of the hill that later became the site of Fort Yuma. After a few months, the soldiers moved to Fort Yuma Hill, where the former Spanish mission settlement of Purisima Concepción had been built. In June 1851, once peace was restored in the region, Camp Yuma was abandoned because of the difficulty of supplying the troops with adequate provisions.

But in late 1851, a general Indian resistance movement along the lower Colorado River was organized, a crusade that included the Quechans, in an attempt to evict all non-Indian intruders. The Quechan chief maintained that the white man and the Indian simply could not live together. Whites, as in the past, could pass through Quechan land, and rest and feed their horses and mules, but they could not build boats and ferry the river. That was the domain of the Indian, and this was, after all, Quechan land.[35] Nevertheless, whites were determined to use the Yuma crossing to their benefit, while the Quechans were just as determined to rid themselves of the foreign intruders. Because of the Indian discord, Major Heintzelman was ordered back to the region to reoccupy the ruins of Camp Yuma, which in a few months would be renamed Fort Yuma, and a military reservation of about twelve square miles was established that straddled the Colorado River at the Yuma crossing. It was now determined that the garrison could be supplied by water transport, and Heintzelman returned at the end of February 1852, as soon as supplies were known to be coming up the river.[36]

Quechan Submission

After an attack by a war party of Quechans on a detachment of Heintzelman's troops who had been sent to the mouth of the river to help organize delivery of the supplies, Heintzelman ordered his men to embark on an aggressive series of campaigns against the Quechans.[37] U.S. troops launched a scorched-earth policy, burning villages and destroying the food supplies of the Indians in order to bring them to submission, and within one year of the uprising, in October 1852, Major Heintzelman ratified a treaty of peace with the Quechans. Coincidentally, another factor that contributed to the defeat of the Quechans was their ongoing intertribal warfare with the Cocopas. The culmination of the Quechan-Cocopa war came in 1852 when the Cocopas carried out an unprovoked attack on the Quechans, killing, according to Heintzelman, eight men and thirty-three women and children. This incident occurred sixty miles south of Fort Yuma, and, because of the Cocopa episode, from that time on the Quechans found it expedient to live near the military post.[38]

It was Heintzelman's belief that it was only a matter of time before "the vices of contact with whites will cause them [Quechans] to dwindle rapidly away, and another race soon occupy their places."[39] Heintzelman's observation was only partially accurate. Although most of Quechan land near the Colorado-Gila confluence came to be occupied by non-Indians, and the Quechans' numbers declined significantly after their defeat, their strength and determination to overcome adversity have allowed them to survive as a tribe into the present.

In 1865, the Fort Mohave Indian Reservation, comprising 75,000 acres near present-day Needles, and the Colorado River Indian Reservation, encompassing nearly 240,000 acres upriver from present-day Blythe, were set aside for several Colorado River tribes who would go there to live, but both were located far beyond the Quechans' traditional homeland at the Colorado-Gila confluence; hence, the Quechans refused to move to these reservations. Most tribal members instead remained in the shadow of Fort Yuma, and continued to utilize the annual overflow of the Colorado River to practice their traditional subsistence patterns, while also performing menial services for non-Indians who lived in Arizona City, later to be renamed Yuma.

Arizona City was founded in 1864, replacing a settlement known as Colorado City established ten years earlier, the latter being washed away by a major flood in 1862. The small riverside community soon became the distribution hub for supplies shipped via the Colorado River for the military posts and mining camps in the region. Steamboating, in fact, began on the Colorado in December 1852 as a result of the establishment of Fort Yuma.[40] Supplies were sent by ocean steamers from San Francisco to the mouth of the river, where they were transferred to river steamers and delivered to towns located along the river. From these points the goods were hauled to their destinations in mule teams. But the shifting channels and great variability of flow of the Colorado made navigation on the river extraordinarily difficult and dangerous, and steamboats were often grounded and lost. Nevertheless, the desert barrier to overland transit to the east and west continued to make the river route from the Gulf of California useful until 1877, when the Southern Pacific Railroad reached Yuma from the West Coast.

In 1871 the seat of Yuma County, one of the four original counties into which Arizona Territory was divided, was transferred to Arizona City from the waning mining town of La Paz. Two years later, in 1873, the name of Arizona City was changed to Yuma. In September 1877, the rails of the South-

FIG. 2.2. Ferry at Yuma Crossing with Southern Pacific Railroad bridge in background, April 5, 1909. (National Archives photo, courtesy of the Yuma Area Office)

ern Pacific Railroad were laid across a swing bridge spanning the Colorado River connecting Yuma with the West Coast, and in 1880 rail connections to the East across southern Arizona were completed. Whereas ferry service, provided by non-Indians, continued long after the completion of the railroad bridge (fig. 2.2), steamboating on the river declined significantly after the arrival of the railroad, since the east-west flow of traffic was far more essential to the needs of the Southwest than the more limited north-south flow offered by the river.[41]

Meanwhile, the military regime at Fort Yuma seemed content merely to ensure that the Indians kept the peace, since there were no apparent plans for Quechan advancement during the three decades that the post remained occupied, except for two attempts to establish a training school for Quechan children that most parents refused to allow their children to attend.[42] By the early 1880s the behavior of the Indians had become so docile that it was decided to disband the military detachment at Fort Yuma. Coincident with the closing of Fort Yuma, in July 1883, President Arthur issued an Executive Order setting aside a sizable piece of land on the east bank of the Colorado River extending north of the Colorado-Gila confluence for a Quechan reservation. The Quechans were unhappy with this location, having traditionally resided, in historic times at least, on the bottomlands west of the Colorado River, near the Gila confluence. They subsequently urged the superintendent of the Colorado River Indian Agency to intercede with the president on their behalf to have the boundaries encompassing their reservation drawn on the opposite side of the river. The president acceded to the change on January 9, 1884, when a second Executive Order canceled the first, substituting a significantly smaller Fort Yuma Indian Reservation on the California side of the river, one that encompassed the area included in the former Fort Yuma Military Reservation (map 2.4). Both Executive Orders provided that the reservation was to be for the "Yuma and such other Indians as the Secretary of the Interior might see fit to settle thereon," although no other tribe was ever resettled on the reservation.[43] These presidential directives exemplify the arbitrary nature by which Indian reservations and their boundaries often came to be established.

Unlike most eastern tribes, the Quechans were not pushed out of their home territory by the pressures of the expanding frontier, largely because the overwhelming majority of land in the arid Southwest held little interest for white settlers. But fertile alluvial lands found in the Colorado and Gila river bottoms represented an exception to the general rule, and in the 1870s farmers and stock raisers began to file land entries near the Colorado-Gila confluence.[44] The pressures of permanent white settlers in the region no doubt helped persuade the federal government to place precise limits on Quechan land. Restricting Indians to delimited areas meant that potential hostilities on the frontier could be controlled, making way for settlement by non-Indian pioneers. The new reservation encompassed an area amounting to fifty-four thousand acres divided between a strip of overflow land flanking the river opposite Yuma upon which the Quechans could continue to practice flood recession farming, as well as some higher and drier mesa land. But the Quechans were not necessarily confined to the limits of the reservation, for they were still generally free to wander where they chose, working odd jobs in Yuma or on the nearby farms and ranches.[45]

Meanwhile, the name of Fort Yuma Hill was changed to Indian Hill when reservation headquarters were established in buildings formerly occupied by the military. This turn of events symbolized the transfer of administrative responsibility for the Quechans from the military to the Office of Indian Affairs. From a proud, self-reliant, and self-sufficient people, the tribe had now been reduced to one dependent on government policies for its survival.

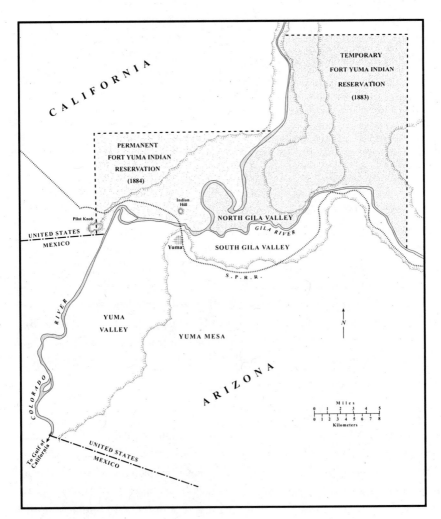

MAP 2.4. Boundaries of temporary (1883) and permanent (1884) Fort Yuma
Indian Reservation. (Year 1883 boundaries based on Executive Order, July 6, 1883,
49th Cong., 2d sess., House Doc. 1, pt. 5, 531–32)

The Quechans had been conquered and subdued; their population had declined from about seventeen hundred in the 1850s when Fort Yuma was
established to approximately nine hundred early in the twentieth century.[46]
Likewise, Quechan land was a mere fraction of its former self. The way had
now been cleared for the irrigation frontier and non-Indian settlement to
advance into the region, a process that in time would cause a further reduction in the limits of Quechan land.

3

Early Irrigation Ventures

The beginning of irrigation near the Colorado-Gila confluence was associated with the early overland stage lines that used the Gila Valley to cross the desert of Southwest Arizona. The first stage line in the region was organized in 1857 and was operated twice monthly between Fort Yuma and Sacaton, a distance of about two hundred miles.[1] It was not until the early 1870s, however, that irrigation of any importance developed in the region. At that time, irrigation was being practiced at several of the stage stations, and one contemporary observer noted that cultivation had begun in the Redondo district, near the junction of the Gila and Colorado rivers.[2] The Redondo district eventually became the largest of these early scattered irrigation developments and represented the first major attempt at irrigation near the Gila and Colorado confluence.

The Gila-Colorado Irrigation Frontier

Jose Maria Redondo, like thousands of other Sonorans, left his home in 1849 to seek his fortune in California's gold country. Although Redondo failed to strike it rich in California, during his ten-year residence there he gained considerable knowledge concerning the details of mining and gold trading.[3] In 1861, while operating a ferry and conducting some mining business on the Colorado several miles upriver from Yuma, he parlayed that knowledge into a fortune. After identifying as gold the particles uncovered by a trapper one hundred miles upriver near the site that would become the settlement of La Paz, Redondo gathered together a contingent of fifty men and essential supplies and headed for the promising new mining district. Their subsequent discoveries led to the "boom of La Paz," which gave the town the distinction of being the largest city in Arizona at its height, and the original seat of Yuma County. The La Paz discoveries also gave Redondo the capital to establish a magnificent rancho, similar to the Spanish ranchos in Mexico and California.

Redondo chose the North Gila Valley at the Colorado-Gila river junction to establish his San Ysidro ranch. By 1871 Redondo's farmlands encompassed nearly two thousand acres, irrigated by a system of canals and ditches totaling twenty-seven miles in length.[4] Brush dams were constructed to divert the waters of the Gila into the canals, while dam and canal construction and maintenance and crop cultivation were carried out by Indian and Mexican laborers numbering two to three thousand. Alfalfa, wheat, barley, oats, and corn, as well as thousands of head of cattle, were raised on the ranch, most of which was sent to the military outposts scattered across southern Arizona. But without title to his land, Redondo was no more than a squatter on the public domain. In 1874 the federal survey of the lower Gila Valley was completed in preparation to opening these floodplain lands to settlement. The threatened loss of its substantial land base caused San Ysidro eventually to wither away.

Nevertheless, it was irrigation ventures along the lower Gila River Valley such as Redondo's that allowed the irrigation frontier to establish a foothold in the Yuma region. In fact, in 1889 a citizens' committee of Yuma County included in its report to the senate committee investigating the irrigation and reclamation of arid lands the assertion that "the Gila river, though second in size to its mighty rival, the Colorado, is destined, for the present, at least, to figure far more prominently in the solution of the problem of redeeming and making valuable to the husbandman the immense bodies of hitherto arid and valueless land tributary to it." The report went on to state that "flourishing ranches in various portions of the [Gila] valley, . . . drawing water from several important canals, amply demonstrate the magnificent results that will ensue should the water supply be rendered permanent . . . through appropriate storage systems." Eugene Trippel, author of the citizens' committee report, also commented elsewhere on the "productive ranches, and extensive irrigation enterprises in progress along the Gila River," while making no mention of similar developments along the Colorado in the vicinity of Yuma.[5]

Although the Gila River was irregular in its flow, and subject to sudden and violent floods, it was a much smaller stream than the Colorado, and therefore easier to divert by individual or cooperative effort. In fact, it was not until the late 1880s that the diversion of the Colorado was considered practicable. One of the earliest and most ambitious proposals involved the diversion of the river's flow into a canal with a heading on the east bank of the river twenty-eight miles above Yuma. The canal was to irrigate twenty-five thousand acres in the Gila Valley before the water would be piped across the Gila River to irrigate seventy thousand acres in the Yuma Valley, and another eighty thousand acres of mesa land. The plan called for the construction of a dam on the Colorado to raise the water twelve feet above the low-water mark, the primary purpose of which was to allow the river's heavy sediment load to settle out in order to clear the river at the point of diversion. A lock was also proposed so as not to impede the river's navigability. The ultimate objective was to carry the canal into Sonora, by concession from the Mexican government, to irrigate additional acreage.[6] Although this venture never went beyond the planning stage, some of its elements were incorporated into later irrigation projects.

Another canal enterprise that never materialized, but nevertheless left an indelible mark in the region, involved the proposed canal of the Colorado River Irrigation Company to divert water from the Colorado near Yuma and carry it to the Colorado Desert, or what would later become known as the Imperial Valley. The idea of irrigating the Imperial Valley using water from the Colorado River originated with Dr. Oliver M. Wozencraft, a physician originally from New Orleans, who resided in San Francisco. From 1853 to 1859, Wozencraft unsuccessfully attempted to raise capital for an irrigation canal system that he had designed that would tap the Colorado River at Pilot Knob near Yuma, traverse south and then west through Mexico, and finally empty into the Imperial Valley. The impenetrable rise of sand hills located west of Yuma necessitated the detour south of the international boundary. In 1859, Wozencraft turned to the California legislature, which adopted a memorial to Congress for a cession to the state of three million acres of federal land in southeastern California to implement the physician's irrigation and reclamation project. In 1862 the plan was brought before the U.S. Congress, and received a favorable report from the House committee reviewing the case, but the measure failed to pass, presumably because of objections of the commissioner of the General Land Office, who viewed the scheme as one that would primarily benefit the private corporation promoting the project.[7]

Wozencraft's plan languished for thirty years, until the early 1890s, when the prospect of reclaiming the Colorado Desert was rekindled by Charles Rockwood, an engineer, who at that time became acquainted with the region and the possibility of its irrigation.[8] The valley was well known to travelers as an uninhabited, grassless, barren plain. But its flat surface and long growing season rendered it particularly attractive to potential irrigators who were convinced of the valley's productiveness if water could be delivered to the region. Rockwood and two associates formed the Colorado River Irrigation Company early in 1890 and began to seek investors in their project. The plan called for the diversion of the Colorado River at the Potholes above Yuma, and installing in the main canal near Pilot Knob a sluiceway that would flush out the silt that escaped being deposited in a short enlarged section of the canal immediately below the diversion point where the velocity of the water would be reduced (map 3.1). Electric-powered hydraulic dredges would periodically remove the material settling out at the point

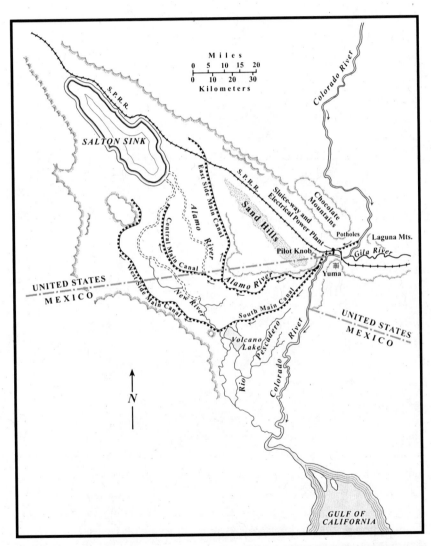

MAP 3.1. Proposed canal system of the Colorado River Irrigation Company. (Based on H. T. Corey, "Irrigation and River Control in the Colorado River Delta," 1232)

of diversion. The Alamo River would then carry the water southward into Mexico, skirting the sand hills, and a set of newly constructed canals would move it into the Imperial Valley, where the water was required for irrigation. Serious investors, however, were not easily convinced that the necessary engineering works could be constructed to irrigate the distant valley. On the other hand, several speculators were found, most of whom were more interested in quick profits than the implementation of the actual project, and by 1893 the Colorado River Irrigation Company was bankrupt.

Seeds of Allotment

But early on, and before the enterprise failed, the promoters had obtained congressional approval for a canal right-of-way through the Fort Yuma Indian Reservation. If water were to be diverted from the Colorado River at the Potholes, the proposed canal would have to pass through the reservation before entering Mexico. In addition to obtaining a right-of-way through the Fort Yuma Reservation, the promoters also wanted to attract non-Indians to the area to boost the proceeds from the irrigation project. Through their urging, an agreement subsequently was struck between the Quechans and the federal government to give five-acre allotments to each member of the tribe so that the remaining irrigable reservation lands could be sold, the proceeds from which would be held for the tribe. Articles I and II of the agreement, signed on December 4, 1893, stated that the Indians "hereby surrender and relinquish to the United States all their right, title, claim, and interest in and to and over" the entire area of their reservation, in return for which each member of the Quechan tribe would be allotted five acres of land.[9] In less than ten years since the founding of their reservation, the land contained within its boundaries was already under siege.

The federal government justified the expropriation of the irrigable portion of the reservation by invoking the Dawes General Allotment Act. Under this measure Indians were to be brought into the mainstream of American society by turning them into successful farmers through the process of allotting them individual plots of ground carved out of their reservation, and opening "surplus" reservation lands to white farmers who would then serve as positive role models. This seemingly well-intentioned federal program, however, was cloaked in a veil of deceit, as it provided a mechanism whereby potentially productive reservation land could be transferred to non-Indian ownership. In the case of the Fort Yuma Indian Reservation, the plan was to provide water from the proposed canal to both the Indian allottees as well as the incoming settlers. The following year Congress passed a statute ratifying this agreement.[10] It gave the Colorado River Irrigation Company three years to begin construction of the canal, or the rights granted the corporation would be forfeited.

The treaty with the Quechans involving the allotment of their reservation was signed by a commission composed of three members appointed by the secretary of the interior as well as a "large majority" of the adult male Indians—203 out of 251.[11] The irrigation company was to benefit from the agreement as a result of white settlement that would be attracted to the canal, in return for which the Indians would have received a perpetual free water right on the lands that they retained. Government documents indicate that the Quechans had initially agreed to the allotment of their reservation in the summer of 1893, before the appointment of the commission.[12] This pact, no doubt initiated by representatives of the Colorado Irrigation Canal Company, was signed by a number of Indians with an X following each of their names, and notarized by Oscar F. Townsend on July 27, 1893.

Townsend is one of several individuals who is often mentioned as a major influence in Yuma's early development. Shortly after arriving in Yuma he became a deputy sheriff, and quickly assumed the role of Yuma County sheriff following the murder of his predecessor by a Quechan. It is said that much of Townsend's work as sheriff involved chasing the Quechans back onto the reservation on the other side of the river.[13] Townsend eventually became Yuma's agent for the Wells Fargo Express Company, a position he held for more than twenty years. However, having dealt with Indian matters since his arrival in Yuma, he remained active over the years in negotiating Indian disputes and in helping the Quechans in other ways.

The Quechans apparently came to recognize Townsend as a person they could trust, regarding him as a sort of "god-father."[14] In fact, one Quechan family adopted the Townsend surname. In connection with the July agreement, Townsend certified "that the foregoing named are members of the Yuma tribe of Indians. . . . That I have been intimately acquainted with said Indians for many years, and that they are anxious to take lands in severalty."[15] This notarized agreement was accompanied by a petition addressed to the president and Congress, containing the same Quechan signatures (also notarized by Townsend), requesting that their reservation be allotted and thrown open to settlement. The petition explained why the Quechans were willing to cede their lands:

We believe if furnished with a small tract of land, with water to irrigate it and with the means of cultivating, we could improve our fortunes to the extent of securing at least all the necessaries of life. We believe from what we hear that if a portion of the land now embraced in our reservation could be thrown open to settlement an irrigating ditch would be built through the reservation. We have noticed that our white neighbors across the river on the Arizona side raise fine crops . . . and they are compelled to cultivate only very small tracts to make a living and more. While with water the soil is fertile, nothing will grow without irrigation, for there is no rain. Hence we want a ditch built so that we can get water and have early and large crops like our white friends. We are willing to give up a large part of our reservation because as it is it is worthless to us, if we can have small tracts set apart for our use. We do not want a great deal of land, for we have noticed that the white man who in this country has small holdings, 5 and 10 acres, and cultivates his land well, is the most successful and the most certain to have what he wants and have it when he wants it.

We believe if our people had small homes of their own, such as they could care for, they would improve rapidly and soon take on the manners of civilization. We only ask that a part of our lands be given us in severalty and under a ditch, and that the proceeds of the sales of the lands of our reservation not appropriated to our use as above be set apart as a fund for our benefit.

We believe our present request to be reasonable and earnestly pray that it may be granted.[16]

In a letter dated September 1, 1893, from the commissioner of Indian affairs to the secretary of the interior, there was acknowledged a request from the Colorado River Irrigation Company for the appointment of a commission to negotiate with the Quechans for the cession of a portion of their lands.[17] Apparently, it was this letter and the earlier petition that inspired the secretary of the interior to appoint a commission on October 19, 1893, to look into the matter of allotting the reservation. Before the commission began its work, the Indians presumably were asked if they had submitted and signed the petition. To this, the Quechans allegedly replied that "it was their petition, that they fully understood it, and were glad the commissioners had come to meet them."[18] They said that they knew of the proposed canal and were glad it was going to be built, as they could then get water year-round and would not have to rely on the overflow of the Colorado River.

The commissioners concluded that it was clear that the matter of the canal had been considered by the Indians, and that from observations of irrigated lands on the opposite side of the river from the reservation they were aware of the benefit the canal would bring them. The commissioners discussed with the Quechans the matter of size of allotments and on what part of the reservation they should be made. It was agreed that all men, women, and children of the Quechan tribe should receive equal allotments, amounting to five acres each, and that it was "for their best interest to locate them in a body and near the headwaters of the canal, or about where the natural flow of the water of the canal appears above the ground."[19] It turns out that this parcel, because of seepage resulting from an unusually high water table, was the most poorly adapted section of the reservation for irrigated agriculture. It was, however, noted by the commission that the land retained by the Indians would be subject to overflow of the Colorado River, as would the bottomlands ceded by them to the government, and that the government should therefore take steps to prevent the overflow by building levees. This, as will be seen later, would only aggravate the seepage problem.

It is apparent that the language of the petition sent to the president and Congress was that of the white man, not the Indian. For example, evidence to be discussed in a later chapter demonstrates that white farmers in the region had no interest in cultivating five- and ten-acre parcels; on the other hand, small parcels allotted the Indians meant more land available to white settlers. One also has to question the genuineness of the passage dealing with the "civilizing" element of allotment. This, of course, was the official government line for carrying out the allotment policy. Without the ability to read, write, or speak English, it is conceivable that the Quechans did not fully comprehend the written word and the course of action that they allegedly were advocating. One can only conclude that the trusted Oscar Townsend helped broker, if not draft, the initial agreement between the canal company and the Quechans, thereby setting in motion the decision by Congress to allot the reservation.

A Yuma resident who had an interest in the welfare of the Quechans as guardian of the children of the deceased parents of one Quechan family wrote the following about the commissioners sent to the region: "I have little faith in that commission or its honesty of purpose. It is generally believed here that the whole thing is a huge swindle in the interest of the said canal company, in whose private interest the Indians are to be ousted." The commissioners responded to this allegation by stating, "They [the Quechans] signed the contract willingly and with as correct a knowledge of what they were doing as a people in their rude state of life could possibly have."[20]

Considering the Quechans' inability to read, speak, and understand English and their characterized "rude state of life," one cannot help but wonder why, although an interpreter was provided them during the negotiations for their land, there was no attorney present protecting their interest in the negotiations.[21] In any case, agreements explained to Indians through interpreters could not be construed with the same degree of precision as contracts between English-speaking Anglos.

A Quechan interpretation of the 1893 agreement, dating from the mid-1930s, disputes the tribe's perceived support of allotment by maintaining that "the will of the tribe was unlawfully controlled by a minority of Indians with the aid of the government officials."[22] It was also alleged that most of the Indians did not understand the interpreter, who spoke in his own language (Mohave), and that with the aid of a number of older Indians, only 148 of the 203 adult male signatures to the agreement were able to be verified, and of those only 77 were residents of the Fort Yuma Indian Reservation. Even though the specific conditions of this agreement were never fulfilled because of the failure of the Colorado River Irrigation Company to proceed with its original plan, the seeds for allotting the reservation and disposing of the surplus land to white farmers had been planted. These seeds would germinate ten years later when Congress, without further investigation, rubber-stamped the allotment of the reservation as a condition of the approval of the Yuma Reclamation Project.

The Imperial Valley Revisited

Not only would allotment appear in somewhat altered form at a later date, but so would the plans of the Colorado River Irrigation Company. After its bankruptcy, Rockwood acquired the engineering records and other data of the defunct corporation, which was then reorganized in 1896 as the California Development Company. The new corporation included Rockwood and one of the other original partners, plus a new promoter of the irrigation project, Anthony Heber. In April 1899, the California Development Company acquired water rights in the Colorado River after posting an appropriation notice under the laws of California for ten thousand cubic feet per second.[23] Since 1893, the year that the Colorado Irrigation Company failed, until the posting of the appropriation notice, Rockwood had sought financial backers for the new project. But toward the end of 1899, the corporation was, like its predecessor, on the brink of financial doom. This time, however, it was saved from bankruptcy by an outside investor, George

Chaffey, a Canadian engineer who had worked on irrigation projects in California before moving to Australia to do the same there. Wozencraft, in his earlier years in California, had approached Chaffey about his scheme for irrigating the Colorado Desert. Chaffey had rejected the idea then on the grounds that, despite its appeal as an engineering enterprise, he believed Anglo farmers were not physically able to work in sustained temperatures of well over one hundred degrees.[24] His Australian experience changed his mind about that. Hence, when Rockwood came to Chaffey with a desperate appeal for financial aid to revive his almost bankrupt California Development Company, Chaffey, attracted by the scope of the project, provided the needed financial infusion and some revised engineering plans that in his opinion would make the scheme more workable.[25] In December 1900 canal construction was begun.

Realizing that the name "Colorado Desert" possessed no appeal to potential settlers, one of Chaffey's first acts was to decree that the region would thenceforth be known as the "Imperial Valley."[26] Chaffey, however, believed that the primary task of the California Development Company should be the construction of the irrigation canal, that the corporation should have nothing to do with the colonization process. A second corporation was therefore formed, the Imperial Land Company, the objective of which was to attract settlers to the Imperial Valley. Anthony Heber soon headed up the Imperial Land Company, and his colonization efforts turned out to be extraordinarily successful.

Heber contracted to do all advertising and colonizing and sell all water stock in return for having exclusive control over townsite selection and a commission of 25 percent on water stock sales. Using government land scrip, the Imperial Land Company obtained ownership of several tracks in various parts of the valley and subdivided them into several townsites, including Calexico, Heber, Imperial, Brawley, and Holtville (map 3.2). Settlers came by the thousands—two thousand by 1903, seven thousand by 1904, and perhaps fourteen thousand by 1905.[27] The overwhelming majority of the new arrivals, approximately 90 percent, acquired their land from the federal government by filing desert land entries that allowed them to enter as much as 320 acres at $1.25 per acre. These entrymen would then be dependent on the California Development Company, and its subsidiary companies, for their supply of water.

Water stock was sold to the settlers for small cash payments and notes payable in five annual installments at 6 percent interest. The first water was

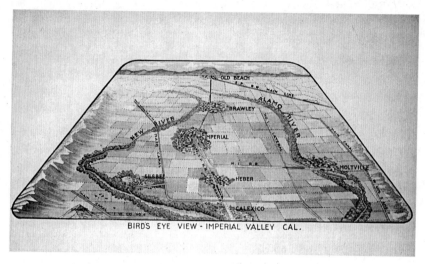

BIRDS EYE VIEW - IMPERIAL VALLEY CAL.

MAP 3.2. Bird's-eye view of the Imperial Valley showing location of townsites. (Reproduced from *Address of Hon. A. H. Heber to the Settlers of Imperial Valley, July 25, 1904*, courtesy of the Water Resources Center, University of California–Berkeley)

turned through the gate in mid-May 1901, but nothing as yet had been done about the distribution canals needed to bring the water to the Imperial Valley farms.[28] Under Chaffey's direction, settlers organized into several mutual water companies that would manage their own canal systems and receive water from the parent company. In June, once the Imperial Land Company had completed contracts with the mutual water companies for the delivery of water, Colorado River water began to trickle toward the budding farms and ranches in the Imperial Valley, but it was not until the following year that the supply of water was able to meet demand. Early in 1904, a forty-mile branch line of the Southern Pacific Railroad was extended into the valley, prompting a spurt in land improvements. In April 1904, nearly 67,000 acres were under irrigation in the Imperial Valley; by the end of the year, approximately 120,000 acres were being cultivated.[29]

The major differences between the defunct Colorado River Irrigation Company's proposed canal and that of the California Development Company were the location of the intake and the means of conducting the water across Mexican territory. Because of financial considerations, a cheaper and quicker method of getting water into the Imperial Valley was devised. Instead of building expensive diversion works, sluiceways, and a settling basin at the Potholes, the revised plan diverted Colorado River water near Pilot Knob on land that Chaffey had secured from an early resident of the area, Hall Hanlon. This site, located just a few hundred feet above the Mexican border, became known as Hanlon Heading. The granitic outcrop of Pilot Knob was the farthest point downriver on the Colorado where a diverting structure could rest on solid rock. And, having reneged on its earlier agreement with the Quechans, this plan would also avoid potential obstacles posed by the Fort Yuma Indian Reservation.

Like its predecessor, however, because of the sand hills that spilled across the Mexican border west of Pilot Knob, the water had to be conveyed through Mexican territory around the southern end of this barrier to reach the Imperial Valley. But since Mexico prohibited the acquisition by foreigners of lands within certain distances of the international boundary, the promoters of the California Development Company organized a subsidiary corporation under the laws of Mexico known as La Sociedad de Riegos y Terrenos de la Baja California (Sociedad Anonym).[30] This company owned the land, the canal, and other properties situated in Mexico. This right, involving the purchase of 100,000 acres plus the bed of the Alamo River where it fell outside this tract, was secured by the promoters of the enterprise from Gen. Guillermo Andrade, Mexican consul at Los Angeles, the original owner of the land. At Hanlon Heading beneath Pilot Knob a seven-mile canal was excavated from the Colorado River across the Mexican boundary to the dry channel of the Alamo River. A short distance above the international border, a temporary wooden headgate, known as the "Chaffey gate," was installed to control the flow of water into the canal.

The revised plan eliminated the longest of the canals of the original plan, the South and West Main Canals, and instead substituted the abandoned channel of the Alamo River, which was graded and reshaped, to convey Colorado River water across Mexican territory. Forty miles west and one mile south of the international boundary, at a place called Sharp's Heading, the canal broke from the bed of the Alamo and fed water back across the border into the canals of the various mutual water companies (see map 3.1). Financial constraints therefore forced the abandonment of the more elaborate plan of diverting the Colorado at the Potholes, where settling basins and sluiceways would have been provided to handle the heavy burden of silt.[31] The elimination of these important features of the project would come to haunt the developers, for when the level of the river fell, the head of the new canal and Chaffey's gate became obstructed by silt deposits. Consequently, a water shortage began to materialize in the Imperial Valley in the fall of 1903, and continued throughout 1904.[32]

Yuma Valley Irrigation and Settlement

While plans were unfolding to divert Colorado River water to the Imperial Valley, developments also were under way that would attempt to irrigate the Yuma Valley, a region with similar characteristics but smaller in areal extent, and one located adjacent to the Colorado River instead of sixty miles distant. Colonization of the Yuma Valley had been delayed until the early twentieth century primarily because of the Algodones Grant that encompassed much of the valley. The land grant included a triangular wedge of land bounded by a diagonal line extending a distance of five leagues in a southwesterly direction from the south side of the Gila River, at what was then its confluence with the Colorado, to the east bank of the Colorado River. Only the east side of the valley was left unencumbered by the grant, and in the late 1870s, after the Southern Pacific's tracks reached Yuma, stockmen began to file desert land entries outside the grant's boundaries, parcels that were used primarily for grazing purposes (map 3.3). These ranchers were interested only in holding their claims for a few years, and had no intention of carrying their entries to patent. Hence, these original entries began to be canceled when the earliest irrigation ventures in the valley attracted a more permanent class of settler.

Coincidentally, the first *patented* entry in the Yuma Valley was filed in the Tucson Land Office by Oscar Townsend in March 1885, who made a half-section desert land entry west of Yuma.[33] But most of the patented entries filed on the valley's east side were made in the 1890s (map 3.4). On average, about three-quarters of the lands in the valley were overflowed each year. The east-side lands were considerably higher than the rest of the valley, and therefore generally not subject to flooding. However, concurrent with these filings, squatters, who also were lured to the region by the nascent irrigation developments, began to take up land within the confines of the Algodones Grant in the heart of the valley. Although the General Land Office did not sanction their colonization efforts, since this land had not officially been thrown open to settlement, they did force litigation involving the authenticity of the Mexican grant.

Proceedings were initiated on February 2, 1892, by the Algodones Land Company, which purportedly obtained the contested land through the pur-

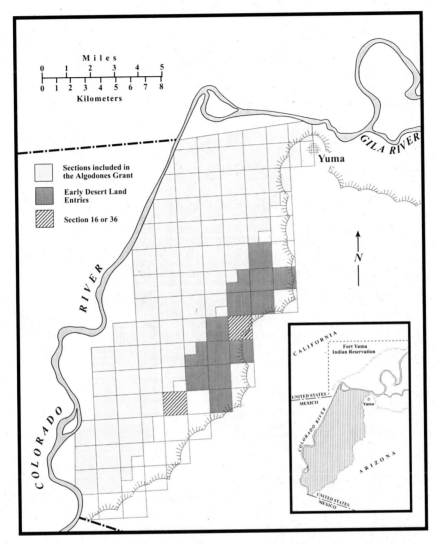

MAP 3.3. Sections included in the Algodones Land Grant and early desert land entries in the Yuma Valley. (Compiled from Department of the Interior, Bureau of Land Management, Tucson District Land Office Tract Books)

chase of the Mexican land grant. The Algodones Land Company was responding to an act approved by Congress a year earlier to provide for the settlement of private land claims in the region, and *El Paso de Los Algodones Grant* was the first case filed before the court. When the grant was originally perfected in 1838, it was done so at a time when the state of Sonora

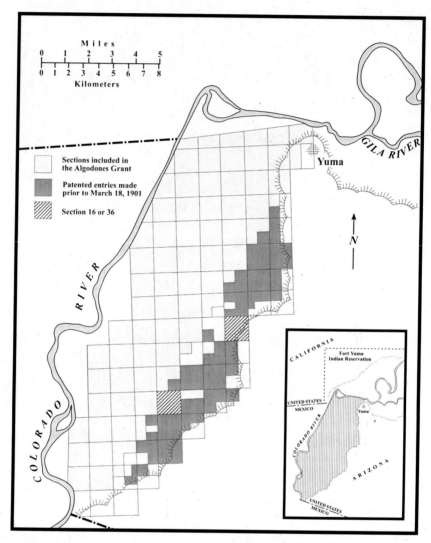

MAP 3.4. Sections included in the Algodones Land Grant and patented entries in the Yuma Valley prior to March 18, 1901. (Compiled from Department of the Interior, Bureau of Land Management, Tucson District Land Office Tract Books)

was in rebellion against the national authority. The court initially rejected the claim on the basis that the signature of the acting state treasurer was not genuine. But at a requested rehearing, the court decided that the signature was genuine after all and that Mexican states had the right to make grants in 1838 on behalf of the national government.[34] Even though the grant had never officially been ratified by Mexico, the Court of Private Land Claims in 1892 confirmed its validity.

Dissatisfied with the decision, the United States appealed to the Supreme Court. Pending ongoing litigation, the Algodones Land Company conveyed the property to Earl B. Coe, and the litigation was revived in his name.[35] Coe was one of the valley's early east-side entrymen, having filed a 480-acre desert land entry late in 1890, of which he patented 160 acres in February 1893.[36] In its hearing of the case, the Supreme Court did not challenge the authenticity of the title papers; instead, in the Court's opinion the principal question was, "Did the officers who made the grant have the power to do so?" The Court noted that even though there was no official claim of a grant from the Mexican government, the grant did state that it was done "in the name of the free and independent sovereign state of Sonora as well as the august Mexican nation." Nevertheless, the Court concluded that because it was acknowledged at the time of the grant that the state of Sonora was in rebellion against Mexico, it and its officers were consequently opponents of the national authority, not its instruments. While declaring independence from Mexico they could not claim to act for it and convey its title, particularly since an 1836 provision in the Mexican constitution had removed the powers of the states to sell lands. Therefore, since the national laws of Mexico were not pursued, the Supreme Court ruled that the "decree of the court of private land claims should be, and it is, reversed."[37]

After being in litigation for more than six years the issue was finally decided on May 23, 1898, in favor of the United States and, by extension, the squatters. Three years later, on March 18, 1901, that portion of the Yuma Valley that was included in the grant was thrown open to entry. Locations were filed quickly to confirm the claims of those who had already begun establishing farms on this land, and in less than three months approximately three-quarters of the patented entries included within the land grant's boundaries had been filed, most of them homestead entries (map 3.5).

More than a decade earlier the first organized attempt to irrigate and grow crops in the Yuma Valley by white settlers had begun.[38] In fact, the first irrigation works in the valley were constructed by the Algodones Irrigation Company in connection with the Algodones Grant.[39] Since the Yuma Valley's terrain slopes gently toward the southwest, in order to cover as much of the valley as possible, the canal's location was placed at the base of Yuma Mesa. It was this development that attracted desert land entrymen to the east-side lands in the early 1890s. Although this part of the valley held the advantage

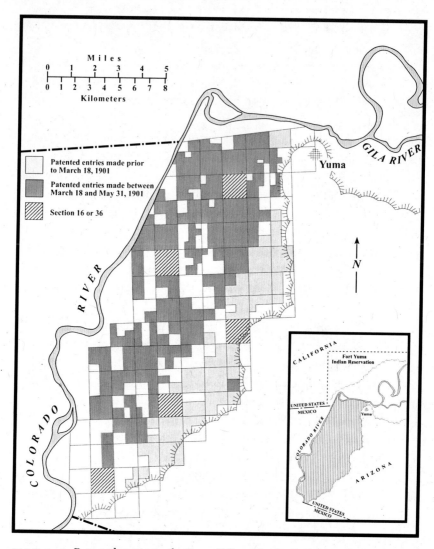

MAP 3.5. Patented entries in the Yuma Valley, March 18, 1901–May 31, 1901. (Compiled from Department of the Interior, Bureau of Land Management, Tucson District Land Office Tract Books)

of being higher and not subject to overflow, water could not readily be brought onto the land by gravity flow; it had to be lifted seven to fifteen feet, depending on the river's stage, by pumping. The pumped water, however, turned out to be very expensive, costing annually from seven to twenty dollars per acre, based on the amount of water used. After the Algodones Land

Company relinquished its rights to the grant, the Algodones irrigation works were purchased by a group of private investors, and the canal came to be known as the Ives Pump Ditch.[40]

The next attempt at irrigation was located near present-day Somerton, about thirteen miles southwest of Yuma, where, in 1895, a group of investors dug several artesian wells and built a large pumping plant. But the texture of the water-bearing sands was too fine to allow the water to flow through them rapidly enough to supply the pumps, and the work was abandoned without any land being reclaimed.[41] Another failed venture was begun in 1897 when a few east-side entrymen and some adjacent squatters on the Algodones Grant joined for the purpose of taking water from the river by means of a gravity canal, known as the American Canal. In 1899, the Colorado Valley Canal and Levee Company absorbed this irrigation system, and in 1900 it was integrated into another private irrigation enterprise, known as the Ludy Canal. To the dismay of many in the Yuma Valley, the Ludy Canal system was to play a fractious role in the development of irrigation in the valley.

Meanwhile, in 1898, a few settlers and squatters in the Yuma Valley co-operatively began the construction of an irrigation system for their own lands. They incorporated under the laws of Arizona as the Farmers Mutual Pump Ditch, and shares of capital stock were issued as water rights. They first completed a pumping plant and ditch, with headworks close to those of the Ives Pump Ditch. As with the Ives Pump Ditch, the Farmers Ditch was intended to irrigate lands too high for gravity irrigation. Soon afterward the Ives Company capitulated to the farmers and discontinued the operation of their plant.[42]

The farmers then embarked on a plan to proceed with the construction of a gravity canal, the Farmers Mutual Gravity Ditch, which was designed to irrigate forty thousand acres of land in the heart of the Yuma Valley. Work on this ditch progressed slowly, however, and it was not until early in 1901 that water was turned into the new canal. Only a small amount of irrigation was carried out from the ditch during its first season of use, however, for when the June high water subsided, the ditch had so filled with sediment that its bottom was above the water level of the river, and no machinery for cleaning the canal was available. The Colorado River carried in its lower reaches one of the heaviest silt loads of any major world river.[43] The settlers therefore quickly realized that keeping the irrigation ditches free of silt and preventing the canal headworks from washing out would be the primary engineering obstacles that they would have to overcome if they were to be successful. There was also the problem of the river's annual overflow that could be checked only with the construction of levees. Nevertheless, in the late 1890s, the settlers, with their two canals, which by now had merged into the Yuma Valley Union Land and Water Company, had absolute control over irrigation matters in the valley.[44]

This situation was soon to change. Sometime in late spring, 1899, J. E. Ludy, a resident of Washington State, came to the Yuma Valley to visit his father, who had taken up residence there. In the fall of 1899, the younger Ludy began to talk of developing an irrigation system that rivaled that of the Farmers Mutual Company. Ludy interested some friends back home in Washington in his plan, who were willing to invest in the construction of the canal. The name of the new corporation was the Irrigation Land and Improvement Company, but it was more widely recognized as the Ludy Canal Company, since Ludy was both chief engineer and promoter of the canal. The Colorado Valley Canal and Levee Company's works were purchased, and surveys were begun that ran parallel to the lines of the Farmers Mutual Gravity Canal, and defection from the ranks of the Farmers Company was encouraged by Ludy. Canal forces were now divided into two groups, and the principal objective of both groups was the irrigation of virtually the same ground.

Although these irrigation ventures more or less duplicated one another, there was an important difference between the two. The Yuma Valley Union Land and Water Company was a cooperative endeavor among the farmers of the valley. The association paid for most of the labor and material furnished by the settlers in capital stock, which became water rights appurtenant to some piece of land under the canal. Farmers under the Ludy system, on the other hand, simply had a license to purchase water; they had no voting power or proprietary interest in the canal, and would pay whatever price the company set for their water. The primary distinction is that in the latter instance water rights were not appurtenant to given parcels of land. This distinction would become an important issue when the Yuma Reclamation Project began to be developed in the region.

When the first water was turned into the Ludy Canal in 1901, it experienced the same kind of misfortune as the Yuma Valley Union Land and Water Company; that is, neither canal was constructed on a grade sufficient to carry the muddy waters of the Colorado very far without silting up. The only means by which the canals could be kept open was by continual dredging, the resources for which neither group possessed. The other source of trou-

FIG. 3.1. Home of wattle-and-daub construction in the Yuma Valley, May 1910. (National Archives photo, courtesy of the Yuma Area Office)

ble, the river's annual overflow that flooded low-lying lands and washed out the canal headings, could be prevented only with the construction of levees. According to O. P. Bondesson, founding president of the Yuma County Water Users Association, the farmers not only struggled with these two difficulties but were also divided into two groups; hence, "the combined strength of the valley could not be used to [maximum] advantage" in an effort to reclaim the region.[45]

Because of these complications, agriculture in the Yuma Valley in the early 1900s was still in its infancy. The typical pioneer farmer lived in a house constructed of wattle and daub (fig. 3.1). At the time, only a few hundred acres in the valley were being cultivated.[46] Alfalfa was the principal crop that supported the majority of farmers; it was sold in the local market. Barley, sorghum, and other field crops were also grown, but in limited quantities. The Quechans from the reservation across the river hired themselves out to dig canals and ditches, and to help with the alfalfa and grain harvests. But the major irrigation works of the valley virtually duplicated one another, and neither was very effective because of the immense problems of controlling and diverting the Colorado River, and because of the lack of solidarity among irrigators. These were the circumstances associated with irrigation in the Yuma region on the eve of congressional approval of the Newlands Reclamation Act in 1902. Needless to say, under the prevailing conditions, the future of irrigation at Yuma did not appear very promising.

4

The Yuma Project

By the time irrigation developments began to unfold in the Yuma Valley, irrigators had already appropriated many of the smaller streams of the arid West. Since it was the smaller streams issuing from the western mountain ranges that were most easily diverted to newly planted fields, at the turn of the twentieth century a fragmented pattern of relatively small irrigated farming districts had become established in the western United States (map 4.1). A pioneer farmer with a team of horses and some ordinary farm tools could single-handedly construct the necessary ditches to water fields located near a small stream, but in order to irrigate more distant farms or higher benchlands cooperative effort among irrigators or the capital of private corporations was required. Hence, the irrigation works of the West fell into two categories: small ditches built by individuals or associations of farmers and larger works constructed as an investment for outside capital and not planned or owned by the irrigators themselves. In the Yuma Valley both types were represented, and each was experiencing major difficulties, primarily because of the unpredictable nature of the Colorado River.

Generally, the irrigation works constructed by the landowners themselves throughout the West were relatively successful endeavors, and most of the land brought under irrigation by the early 1900s had been accomplished in this fashion. These tended to be cooperative but relatively small-scale ventures. The problem encountered by Yuma irrigators was that the farmers who had joined to form the Yuma Valley Union Land and Water Company were attempting to divert a major river without the financial means to do so. On the other hand, the works planned and built with money borrowed from investors intending to sell water rights, such as the rival Irrigation Land and Improvement Company's (Ludy) canal, were almost without exception financial failures. Most of the corporate ventures were engineered and promoted by incompetent or speculatively inclined persons having insufficient capital to build the dams, storage reservoirs, and canals necessary to irrigate large tracts of land.

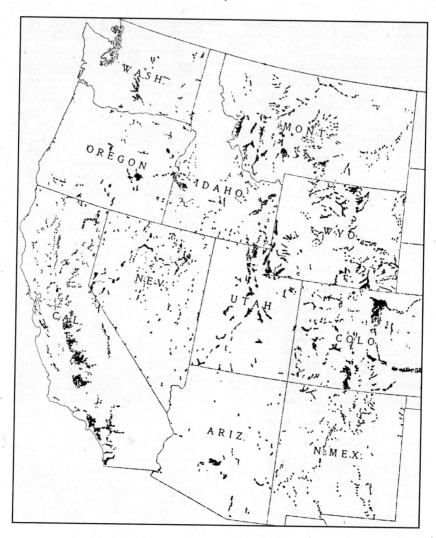

MAP 4.1. Distribution of irrigated lands in the western United States in 1899. (Reproduced from U.S. Bureau of the Census, *Twelfth Census of the United States Taken in the Year 1900: Agriculture—Part II, Crops and Irrigation*, 802)

The primary problem confronting both the farmers and the investors was their inability to keep the Colorado River in its channel during flood stage and to keep the canals from silting up because of the enormous sediment load carried by the river. Ironically, one of the few large-scale private irrigation endeavors in the West that showed more promise than most involved the activities of the California Development Company, which was conducting Colorado River water through the abandoned channel of the Alamo River across the desert of northern Mexico into the Imperial Valley sixty miles away. The grade of the Imperial Canal from its heading on the Colorado River to the Salton Sink was more than three times as steep as that of the river below its heading. This normally was a sufficient gradient to keep the main canal from silting up, and the territory it irrigated was far removed from the river's annual overflow.

The situation that prevailed in the Yuma Valley was metaphorical; it reflected the federal government's concern that irrigation in the West had gone about as far as it could. The lands that could readily be irrigated through individual or cooperative initiative were now occupied, and some form of federal intervention would be required if the remaining irrigable lands were to be populated by small family farmers. Vast chunks of the western public domain were rapidly being taken up by stockmen through fraudulent entries made under the Desert Land Act, while the objectives of the homestead law, to permit the expansion of the country's agricultural area and to provide homes as rapidly as the needs of the people demanded, were not being met. The solution to this dilemma involved the implementation of a national irrigation policy through passage of the Newlands Reclamation Act in 1902.

Colorado River Surveys

The Newlands Reclamation Act provided for the examination, survey, and construction of irrigation works that would enlarge the irrigated acreage in the West, the funds for which would come from the sale of the public lands in the sixteen states and territories of the arid region that would benefit from the new law. Consequently, in October 1902, the Reclamation Service began topographic and hydrographic investigations for the development of irrigation projects along the Colorado River, from about one hundred miles above Needles to the Mexican border. Arthur Powell Davis, nephew of John Wesley Powell and a supervising engineer of the Reclamation Service, proposed the general plan for reclamation of the lands adjoining the Colorado River. According to Donald Worster, Davis's driving ambition was the conquest of the Colorado River. Davis observed that the lower Colorado River valley generally was relatively narrow, but that there were places where water could be taken out to irrigate thousands of acres. Davis recommended that the government construct several high dams across the Colorado, creating a series of large reservoirs, which would utilize the floodwaters of the river. The dams were not intended to interfere with the river's regular flow, since a portion of its flow had already been appropriated by the California Development Company and the smaller irrigation ventures in the Yuma Valley. In order to be able legally to fill the proposed reservoirs, in August 1903, the Reclamation Service filed on the unappropriated waters of the Colorado River in both California and Arizona. This filing acknowledged the fact that other appropriations had already been made on the waters of that stream, for the government's notice clearly stipulated *unappropriated* waters.[1]

Davis suggested that the project located farthest downriver at Yuma should be taken up first, and additional reservoirs created as needed. With a high dam located near Yuma, the original plan also included the possibility of incorporating the Imperial Valley into the project. Presumably, the government initially had no intention of interfering with private corporations or attempting to drive them out of business, for this was clearly denounced in Section 8 of the Reclamation Act. On this issue, however, Reclamation's attitude quickly changed to one embracing the argument that it was unfair "to the communities concerned . . . for the Reclamation Service to step aside in favor of speculative enterprises, especially when they would only partly develop the opportunities."[2] And it was evident, at least to Reclamation Service engineers, that private enterprise on the Colorado could not cope with the problems of reclamation as effectively or as efficiently as the government.[3] For example, chief engineer Newell rationalized that if the job of reclamation in the Imperial Valley were completed by the federal government, it would result in far more benefits to Imperial Valley residents, including the ability to purchase water cheaper than from the private developers and the eventual control and ownership of the irrigation system by landowners after the construction costs had been paid off. The absorption of the Imperial Valley into the federal project also offered a chance to expand significantly the Yuma Project in terms of acreage to be irrigated, and, although the overall cost of the project would be increased, the average per-acre cost would be reduced.[4]

Prior to Davis's Colorado River survey, Joseph B. Lippincott had been hired by the Reclamation Service to serve as supervising engineer for California projects. He immediately wrote Senator Thomas R. Bard from California, who was strategically placed in Congress as chairman of the Senate Committee on Irrigation and Reclamation of Public Lands, and who also served on the Committee on Indian Affairs. Lippincott introduced himself in the letter as representing the Hydrographic Branch of the U.S. Geological Survey in California, and expressed his belief that "one of the greatest opportunities for national irrigation in California consists in the development of the possibilities along the Colorado River, between Needles and Yuma."[5] Lippincott then invited Senator Bard to visit his Los Angeles office when convenient so that he could be apprised of developments, in order to be in a position to provide congressional assistance for the program if he was so inclined. Lippincott during the next two years would rely heavily on Senator Bard's influence in Washington, and Bard turned out to be the key player in obtaining federal approval for the Yuma Project.

Before initiating surveys along the Colorado River, public lands were temporarily withdrawn from entry under the terms of the reclamation law (map 4.2). The law stipulated two categories of withdrawal. *First-form* withdrawals included areas where irrigation works (reservoirs, canals, and so on) would likely be built. No entries of any kind were allowed on these lands. *Second-form* withdrawals included irrigable lands. These lands could be entered, if they had already been proclaimed open to entry, but only under the provisions of the Homestead Act. Although nearly three-quarters of potentially irrigable land in the Yuma Valley had already been alienated by this time, on July 2, 1902, the remaining unentered lands were withdrawn under the first form. This action, however, was quickly amended the following month when they became second-form withdrawals, allowing entries under the Homestead Act.

Meanwhile, a soil survey of the Yuma Valley had already been completed during the winter months of 1901–1902, and in the following year the survey was continued on the California side of the river to cover lands encompassed by the Fort Yuma Indian Reservation. The surveyors concluded that before the natural fertility of the floodplain soils could be fully exploited, the river's annual overflow would have to be controlled. However, it was emphasized that since the water table of the valley rises and falls in concert with the river, leveeing the river would cause it to rise higher in its channel, thereby forcing groundwater too close to the surface in many places for successful farming. In fact, in areas bordering the river, it was cautioned, the groundwater would seep above the surface. For farming to be successful, according to the government soil experts, the installation of an expensive drainage system would be required. Lippincott, a firm believer and strong advocate of the importance of developing the irrigation possibilities of the Colorado, was convinced that no matter how complex the environmental problems associated with reclamation were, they could be conquered.[6]

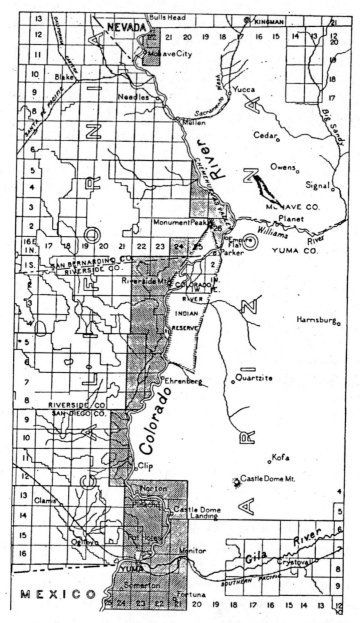

MAP 4.2. Lands temporarily withdrawn along the Colorado River in California and Arizona. (Reproduced from Department of the Interior, U.S. Geological Survey, *First Annual Report of the Reclamation Service From June 17 to December 1, 1902,* 107)

But there was also a human element that complicated matters. It was apparent to Lippincott that if a reclamation project were to be constructed in the vicinity of Yuma, from an economic standpoint it would be necessary to incorporate the Fort Yuma Indian Reservation into the overall plan. Even before the Colorado River surveys were begun in October 1902, the secretary of the interior, working under the assumption that the Quechans would be allotted, ordered a field examination of the reservation so that an irrigation plan for the reservation could be prepared. W. H. Code, an engineer employed by the Indian Service, was placed in charge of the survey, and, in October 1902, he submitted his report to the secretary of the interior.[7] Code suggested that the only satisfactory heading for a canal was at the Potholes, several miles above Yuma, where the river channel was permanently entrenched in bedrock. With the headworks at this location the canal should have sufficient gradient to alleviate the siltation problem that was causing so much trouble for the irrigators in the Yuma Valley.

Surprisingly, Code's recommendations involving farming on the reservation assumed that levees would not be constructed to contain the river's overflow. He estimated that about seven thousand acres of land on the reservation were sufficiently elevated to allow for constant alfalfa production from a canal diverting water at the Potholes, although small levees would be necessary to keep out the highest floodwaters. The remaining eight or nine thousand acres of low land could be used for growing corn, pumpkins, and melons in the traditional fashion after the floodwaters subsided. But this land could produce even more abundant returns if given a supplemental supply of water from the canal. Code went on to state that according to John Spear, the superintendent of the Yuma Reservation, no more than two hundred families would avail themselves of land under the proposed canal, and that forty acres would be more than ample for the needs of each family. Hence, in his judgment, approximately eight thousand acres of reservation land would fulfill their requirements.

Following the receipt of Code's report, the secretary of the interior requested an opinion from the commissioner of Indian affairs regarding the allotment of the Fort Yuma Indian Reservation. In response, the commissioner quoted from Spear's recent annual report. Spear wrote, "I am more convinced every day that if their land were irrigated and allotted, the Indians would work it profitably and successfully. There is no other hope for them. The Yuma have reached that stage where it is impossible for them to remain. With irrigation and allotment, which means employment, I believe they will go forward. [Otherwise] . . . idleness will turn them into loafers, gamblers and drunkards. Now is the time to work for the Yumas, not after these habits are formed." Based on Spear's comments, the commissioner concluded, "The allotment of these lands appears to be desirable." He seemed convinced that if the government built a canal and the Indians were allotted their land, they would become permanently self-supporting. But in hindsight, the commissioner became concerned about the possibility that his overzealous support of allotment might be misinterpreted; that is, he feared that the government might begin to move too quickly on the disposal of surplus lands in the reservation, and in a follow-up statement to the secretary of the interior, he clarified his views. He emphasized that "no definite steps should be taken looking to the disposition of any of the surplus lands of the reservation, either irrigable or non-irrigable, until provision shall have been made to furnish the Indians with allotments of irrigated lands."[8] Either way, the allotment of the Fort Yuma Indian Reservation, at least in principle, had the support of the commissioner of Indian affairs.

Impediments to Reclamation

Despite the complication involving the Fort Yuma Indian Reservation, by the summer of 1903 Lippincott concurred with Davis's recommendation that the region around Yuma was the most desirable locality to begin construction on the Colorado River. During the following winter months, therefore, Reclamation Service engineers conducted more detailed surveys for the irrigation works that would be constructed near Yuma. Bedrock, however, was nowhere to be found at sufficient depth to provide a proper foundation for the construction of the seventy-foot-high dam proposed by Davis. But since the greatest flow of the Colorado nearly always came in summer when the greatest demand by irrigators existed, storage was not absolutely necessary.[9] Hence, a low diversion weir dam was substituted for the proposed high dam at Yuma. In March 1904, Lippincott and the reclamation engineer who would become the first project engineer at Yuma, Homer Hamlin, jointly submitted a preliminary plan for the Yuma Project. The total cost of the project was estimated to be $2.7 million and the net irrigable area was given at 76,966 acres, for a per acre cost of $35.10. The following month a board of six consulting engineers headed by A. P. Davis examined the plans and estimates submitted by Lippincott and Hamlin, and recommended approval with minor adjustments.[10] The board then urged that Lippincott be instructed to proceed with the preparation of detailed plans and specifications for the project, that the secretary of the interior give his general approval to the plan, and that $3 million be set aside for construction. However, embedded in the board's report was an accounting of several obstacles that would have to be overcome before the project would actually be feasible.

First, although the flow of the Colorado was sufficient for all requirements of the Yuma Project, because of its status as a navigable stream, the legality of the diversion of the river had to be authorized by Congress. Second, although the lands included in the proposed project were generally fertile, those on the California side were included entirely in the Fort Yuma Indian Reservation. The board regarded these lands as essential to the successful completion of the project and knew legislation would be required to guarantee that they bear their share of the cost of construction. Third, the lands on the Arizona side were mainly in private ownership, and the project would not be feasible unless practically all landowners contributed to the total cost. It was therefore imperative that landowners in the Yuma Valley execute liens on their lands for the return of the cost of reclamation and that those holding more than 160 acres agree to dispose of the excess in compliance with the provisions of the Reclamation Act. The Yuma Valley settlers had already been informed that the government would deal with them only as a corporate body, and not individually as private landowners. In fact, in November 1903, at the urging of Reclamation officials, an organization known as the Yuma County Water Users Association had already been formally incorporated for the purpose of dealing en masse with the government. The board concluded that only after these impediments were successfully dealt with should construction of the Yuma Project begin.

The Fiction of Navigability

The navigability of the Colorado River was sheer fabrication by the Reclamation Service, a maneuver that involved political "sleight-of-hand" designed to eliminate private competition in the region. The low diversion dam now contemplated at Yuma, in lieu of a high dam and large storage reservoir, represented a serious deterrent to the project, for it meant that no longer was the appropriation by the Reclamation Service to be made from the surplus water stored in the reservoir; instead, it would be the actual flow of the river. This would encroach on the prior appropriations of the Colorado Development Company and the smaller canal enterprises in the Yuma

Valley. The politically expedient way to handle this predicament was to have the Colorado River redeclared a navigable stream, thus making the prior appropriations illegal. That is, if the Colorado could be considered navigable, then its flow was not subject to the appropriation laws of California and Arizona, and hence the settlers under the existing canals had no right to use the water from those systems.

The Colorado River, for a distance of twenty miles beginning at a point about eight miles below Yuma, marks the boundary between the United States and Mexico. The navigability of that part of the river delimiting the international boundary was originally recognized when Mexico and the United States entered into the Treaty of Guadalupe Hidalgo in 1848.[11] A subsequent treaty in 1853 canceled this provision, but guaranteed to the United States the free and uninterrupted passage of vessels and citizens where the river forms the common boundary between the two countries. In spite of the latter agreement, there had been practically no commerce on the river below Yuma since the Southern Pacific Railroad began to serve the region in 1877. U.S. Army engineers subsequently substantiated the river's minimal importance for navigational purposes by reporting, on several occasions, that the navigational interests were not sufficient to justify any expenditure for river improvement.[12] Nevertheless, if the Colorado could somehow retain its navigable status, then prior appropriations of the river's flow could be declared illegal.

The Department of the Interior, however, had already incidentally recognized the legality of water rights of the various canal companies when it authorized canal rights-of-way on public domain land. Likewise, when it patented desert land claims, Interior was indirectly validating water diversions, since desert land entrymen were required to state in their applications where their water for irrigation purposes would come from. When these appropriations were made, however, the legality of the diversions from the Colorado was not an issue. Now, because of the presumed navigability of the Colorado, all diversions, including the government's, would require congressional permission. This meant that the California Development Company and the Yuma canal companies would have to seek federal permission to continue to take water from the river. Congressional refusal would be one means for the government to complete its own irrigation scheme and take over the irrigation systems in both regions.

But the navigability of the Colorado would not be an easy issue to prove. Late in 1901, in anticipation of the passage of the Reclamation Act, an investigation was conducted by the State Department to ascertain the river's navigability. It concluded that no vessels at that time were using the Colorado River below Yuma.[13] And in July 1902, not long after the approval of the Reclamation Act, an International Boundary Commission was established to determine the extent of navigability of the Colorado since future dams placed across the river in the United States could impact treaties with Mexico. The commission found that the swift shoal waters and the shallow depth over bars in the river, together with a thirty-foot tidal bore at the river's mouth, had resulted in practically no commerce on the river. In fact, since 1877, when the Southern Pacific completed its track to Yuma, only five or six trips by boat had been made below Yuma, and all but one occurred prior to 1898. The commission therefore concluded, "The value of water to the two Republics who use it for irrigation, will be many hundredsfold greater than when let run useless to the sea on the chance that the owners of some small steamer might wish to make sporadic trips."[14]

Fortuitously, however, from the standpoint of the Reclamation Service, a company called the Mexican-Colorado Navigation Company was organized early in 1902 to operate a line of steamships to transport mined ore from southern Nevada down the Colorado, and on to Guaymus, Mexico. The very existence of this firm allowed the Reclamation Service to show actual evidence of the *potential* use of the Colorado for navigational purposes.[15] Apparently, this was enough to convince Congress that the Colorado should be officially designated a navigable stream.

A year later, M. C. Burch, representing the Justice Department, prepared a special report concerning use of the Colorado for irrigation and navigational purposes. Burch ascertained that no treaty rights with Mexico were being violated by the California Development Company, or other canal companies, and that the water extracted from the river had not lessened the navigability of the river. On the other hand, even though Burch acknowledged in his report that the possible use of the river for navigation was of small importance in comparison to its use for irrigation, according to Congress, the river nevertheless did fall under the jurisdiction of a navigable stream. Therefore, from a technical standpoint the ongoing water diversions were illegal. The attorney general, however, doubted the soundness of the government's case, and ultimately the Reclamation Service failed in its attempt to prove that the private irrigation developers on the lower Colorado River were guilty of the misuse of a government-controlled stream.[16] Meanwhile, however, the government's ploy bought enough time to at least partially achieve the purpose for which it was devised.

Lippincott realized that the government's jurisdiction over the Colorado as a navigable river stood on shaky legal grounds. In mid-March 1904, he wrote Senator Bard explaining that the Colorado River legally could not be diverted by the government or anyone else without the consent of Congress, and that it was of vital importance to the irrigation movement in California that authorization be granted the government during the current session of Congress. Lippincott apprised Bard of the fact that Congressman Daniels, on behalf of the California Development Company, had introduced a bill in January that would declare the Colorado a nonnavigable stream, thereby making their diversion legal. This appropriation represented more than the mean annual flow of the river. The Daniels Bill argued that "the water of the Colorado River for the irrigation of the arid land that may be irrigated therefrom is hereby declared to be of greater public use and benefit than for navigation, and the diversion of water from said river . . . [which] may in future be made, for irrigation purposes, in accordance with the laws of the respective States and Territories in which such diversion has been or may be made, is hereby legalized and made lawful."[17]

Lippincott feared that if Congress approved the Daniels Bill, it would place the claim of the California Development Company to the river's flow prior to that of the government's. Lippincott professed to Senator Bard, "I am reviewing these matters to you unofficially and informally . . . with a hope that you will urgently advance our efforts to get Congressional sanction for the use of this river by the Reclamation Service through this session of Congress."[18]

Lippincott had a sympathetic ear, for Senator Bard was no friend of private enterprise where irrigation was concerned. In a previously unpublished statement attributed to Bard that appeared in the *Third Annual Report of the Reclamation Service*, the senator expressed his feelings on the subject of national vis-à-vis private irrigation developments:

It is recognized that the primary purpose of the Reclamation Act is to utilize the public domain by means of irrigation and make it available for occupation by settlers. By keeping the obvious intent of the law in mind it will be possible to solve many of the difficult questions which arise when private irrigation enterprises interfere with Government projects. It is clearly the duty of the Reclamation Service to investigate and determine whether a given private enterprise is competent to utilize the irrigation resources to the fullest extent and whether the private works are of a permanent character. It is entirely proper for the Service to refuse to abandon its own projects until satisfactory guaranties [sic] are given that these requirements will be provided.[19]

In other words, Bard felt that the government should not defer to private enterprise simply because the latter had made its appropriations and constructed its irrigation works first.

In order to cover all his political bases, Lippincott also wrote George Pardee, governor of California, providing the governor with an update on "some matters relative to the use of the Colorado River under the Reclamation Act." Lippincott informed Pardee that the secretary of the interior, the Indian commissioner, and the chief engineer of the Geological Survey had each written letters urging Congress to provide for the diversion of the Colorado River, and to authorize the opening of the Fort Yuma Indian Reservation under the general provisions of the Reclamation Act. Lippincott emphasized that if specific authorization from Congress to make a legal diversion was not immediately forthcoming, the project could be delayed for the foreseeable future. He also pointed out that certain corporations, namely, the California Development Company, were already making diversions from the Colorado, and it was his opinion that "the prompt utilization of the water in California and Arizona is the very best protection that we can give to our interests and our rights to that stream."[20] The underlying motive for Lippincott's letter to Pardee, however, stemmed from the fact that California's other senator, George C. Perkins, had written to the secretary of the interior pleading for his support in the ongoing private irrigation developments of the Imperial Valley. Lippincott therefore urged the governor to take whatever action he deemed appropriate to assist the government in its campaign.

Meanwhile, internal differences among the operators of the California Development Company had caused Chaffey to sell his interest and leave the corporation; Rockwood assumed the position of chief engineer, while Anthony Heber became president of both the Imperial Land Company and the California Development Company.[21] When Lippincott originally questioned the right of the California Development Company to divert water from the Colorado River, Heber lobbied for the Daniels Bill before the Committees on Irrigation in both the House and the Senate.[22] He argued that the river's appropriation by the company was prior to the enactment of the Reclamation Act, and prior to the filing and the appropriation made by the government under said act, and that the good faith of the company had been shown in proceeding to make use of its appropriation, thereby establishing the validity of its right to take and use the water. The Daniels Bill, on the other hand, was bitterly opposed by William Ellsworth Smythe. Smythe, who as we have seen was prominent in the reclamation movement, claimed to represent the settlers in the Imperial Valley, who he said were eager to be included in the Yuma Project.[23]

On April 8, 1904, the acting attorney general was called before the Senate Committee on Irrigation to state his opinion on the navigability question. In his view, because of provisions included in the Treaty of Guadalupe Hidalgo, the Gadsden Purchase, and the boundary treaty with Mexico, and because of the important irrigation projects proposed by the government, he seriously doubted "the wisdom of a surrender by Congress at this time of all control over the waters of the Colorado River."[24] Senator Bard, who chaired the committee, sided with Lippincott, Smythe, and the attorney general. He subsequently threw the weight of his influence against the Daniels Bill, and the bill died in committee. The committee's rationale for not proceeding was that if the bill were passed, it not only would violate treaties with Mexico but would also create a monopoly and place farmers at the mercy of corporation owners.

As a concession to the California Development Company, a joint resolution of Congress was subsequently approved directing the secretary of the interior to appoint an investigating committee to ascertain what, if any, legislation would be necessary to grant and confirm to private enterprise and appropriators rights to divert water from the Colorado River for irrigation purposes. The committee, however, could not have been more biased in favor of the government, for it was chaired by chief engineer Newell and included supervising engineer Lippincott along with several other Reclamation Service engineers. According to Heber, the delegation paid an initial visit to the Imperial Valley, spending only one day, in the summer of 1904, then returned again in the fall, but presumably its members were more interested in determining the sentiment of Imperial Valley residents regarding their inclusion in the general irrigation works to be constructed at Yuma than in ascertaining the kind of legislation necessary to make their appropriations legal.[25]

After Senator Bard oversaw the defeat of the Daniels Bill, he then exerted his influence as a member of the Committee on Indian Affairs by introducing a bill that specifically authorized the secretary of the interior to make such diversions in both Arizona and California as might be necessary for the irrigation of lands adjacent to the river. Both the Fort Yuma and the Colorado River Indian reservations were located on such land. In order to gain prompt approval, Bard had the following clause attached to the annual Indian appropriations bill, approved April 21, 1904: "The Secretary of the Interior is hereby authorized to divert the waters of the Colorado River and to reclaim, utilize, and dispose of any lands in said reservations which may be irrigable by such works in like manner as though the same were a part of the public domain: *Provided,* that there shall be reserved for and allotted to each of the Indians belonging on the said reservations five acres of the irrigable lands. The remainder of the lands irrigable in said reservations shall be disposed of to settlers under the provisions of the Reclamation Act."[26] This last-minute attachment to the bill carried with it enormous ramifications, for, on the one hand, it sanctioned the government's diversion of the Colorado River, and, on the other hand, it authorized the allotment of adjacent reservation lands. With the exception of one remaining potential obstacle, the way had now been cleared for the construction of the Yuma Project.

The diversion of the Colorado meant that an agreement with Mexico over placing a dam across the river might have to be worked out. Lippincott brought up this matter in a letter to Newell in April 1904, in which he informed Newell that the Justice Department's Burch had advised him, "We have the power, as far as treaties with Mexico were concerned, to divert the water."[27] An agreement with Mexico was therefore not pursued. Government officials reasoned that since the headwaters of the Colorado and most of its drainage were found in the United States, it was not necessary to consult with Mexico about the dam. A selfish policy toward the river that reflected the national interest of the United States was therefore pursued.[28] Once all political barriers standing in the way of the Yuma Project had been overcome, the secretary of the interior gave the go-ahead for its construction.

Yuma Project Components

Authorization for the Yuma Reclamation Project came on May 10, 1904; it was the fifth to be sanctioned by the Reclamation Service, and the first to be constructed on the Colorado River. The immediate task now was to encourage the farmers of the Yuma County Water Users Association to endorse the project by placing a lien on their property for repayment of the project's construction charges. A week prior to the secretary of the interior's approval, a letter had already been sent to him from the president of the Yuma County Water Users Association conveying "the earnest desire and request of this Association . . . for the construction of irrigation works on the lower Colorado River."[29] The water users requested that Yuma Valley settlers be provided a statement regarding the character and cost of the proposed works, as well as more information on the manner in which the association should proceed to secure construction.

Accompanying the letter were two petitions, one signed by members of the board of governors and the other signed by "a large number of leading farmers and landowners of this section" urging the construction of irrigation works on the lower Colorado River. In a letter signed by chief engineer Newell, the Reclamation Service addressed the concerns of the Water Users Association in less than a week following approval of the Yuma Project. Newell's letter was actually drafted by Lippincott a month earlier, who apparently anticipated such a request, for in his cover letter to Newell he stated that "the Board, as well as myself, are of the opinion that it can hardly be expected that these farmers will agree to sign contracts for their lands without having some information as to what they are getting for their money. Therefore, it is only fair that they should be given a general statement concerning the plans of this project. This statement, I have endeavored to present for you to sign, in a letter addressed to the Yuma [County] Water Users Association."[30]

Newell's letter of reply to the association outlined all aspects of the project including the areas to be irrigated; the location of headworks and canals, levees, and drainage systems; and, most important, the cost of the entire enterprise. Newell pointed out that the Yuma Project was to be located on both sides of the Colorado River near Yuma and that the irrigable valley lands under the gravity canal system totaled 86,700 acres, of which 73,100 acres were in Arizona. A weir-type dam would be built about ten miles above Yuma at Laguna, also know as the Potholes, the last narrow point on the Colorado where a dam could be placed (map 4.3). The type of weir to be constructed was similar to that developed by British engineers in their irrigation work in India and improved and used later on the Nile, where several had been operating for many years under conditions similar to those found on the Colorado River. Laguna Dam would not only be the first irrigation structure built on the Colorado River but also be the first of its kind in the United States. For its entire length of slightly less than one mile it would rest on alluvial river deposits, and only at the ends would it be anchored to bedrock. It is important to note, however, that although the dam enabled the river level to be raised a few feet so that irrigation water could be diverted at times of low water, it had almost no storage capacity in its reservoir and therefore afforded no protection against floods on the lower Colorado.

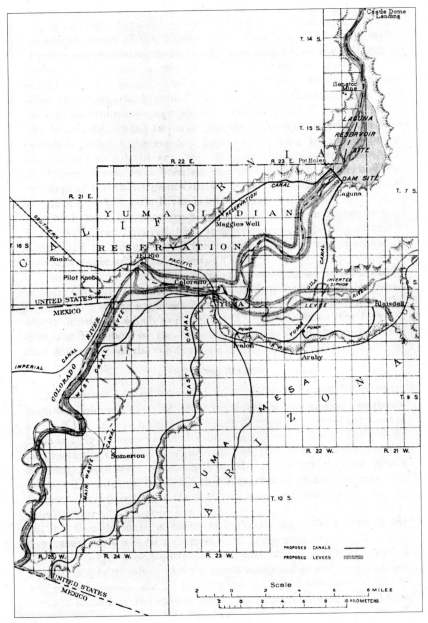

MAP 4·3. Proposed Yuma Project, Arizona-California. (Reproduced from Department of the Interior, U.S. Geological Survey, *Third Annual Report of the Reclamation Service, 1903–4* plate 12)

Realizing that the heavy sediment load of the Colorado was the worst enemy of the potential irrigator through silting of the canals, the most difficult feature of the undertaking would be handling this problem. The dam would raise the low-water level of the river about ten feet and create a ten-mile-long settling basin of relatively quiet water, so that most of the sediments carried by the river would be dropped before entering the canals. Water would then be diverted from the reservoir into sluiceways on either side of the dam where even more silt would be deposited. At the canal intakes, the diverted water would be skimmed from the surface of each sluiceway by the canal headgates, allowing water with a reduced burden of silt to enter the canals. When filled with sediment, sluice gates would be opened so that the alluvial material could be flushed from the sluiceways back into the river below the intakes. As a further precaution, the first three thousand feet of canal on each side of the river would be of such size as to cause the water to move through them very slowly, thereby forming additional settling basins that could be sluiced periodically. At the lower end of the settling basins the canals proper would begin. They would be designed so that the water in them would flow at such a velocity as to retain whatever silt was still carried by the water. This material, containing important fertilizing properties, would then be deposited on land subsequently brought under irrigation.

The Reservation Canal would irrigate the Yuma Indian Reservation lands on the California side, while the Yuma Canal would convey water to the Arizona side across the North Gila Valley, through an inverted siphon constructed beneath the bed of the Gila River to the South Gila Valley, and from there to the Yuma Valley, where it would divide into two branches, the East Side and West Side canals (see map 4.3). (Because of the unpredictable nature of the Gila River during flood stage, the siphon drop would later be shifted to the Colorado River.) A pumping plant on the Yuma Canal at the base of the Yuma Mesa would lift water to the mesa top, but the mesa lands would be developed at a later date and therefore were not included in the original project costs.

Because of the annual summer rise of the Colorado River and the occasional severe floods of the Gila River, levees would be constructed to protect the overflow lands. Drainage would also be necessary due to the flatness of the irrigable lands and the high water table. A main drainage canal therefore would be run through the central portion of the areas to be irrigated. The cost of the entire project, excluding the mesa lands, would result in a levy on landowners of about thirty-five dollars per acre irrigated, provided the lands in private ownership on the Arizona side were included in the project. Newell added an important caveat, however, stating that it was possible, as construction work proceeded, the cost would increase, although the estimates were considered to be extremely "liberal." Newell concluded his letter by explaining that the secretary of the interior had reserved three million dollars of the Reclamation fund for the construction of the project, contingent upon the action of the valley's landowners.

Landowners' Approval

The Yuma County Water Users Association authorized the printing of one thousand pamphlets containing both the letter of the president of the association requesting more information and Newell's detailed reply. Then on the morning of May 28, 1904, the pamphlets were distributed on the streets of Yuma. Later in the day a meeting of the Board of Governors of the Water Users Association was held, followed by an open meeting of parties interested in the project. Present at the latter meeting were chief engineer Newell, supervising engineer Lippincott as well as other government engineers, William Smythe, plus nearly all the landowners in the Yuma Valley.[31] Smythe gave the opening address, followed by Newell and the other engineers who explained the main features and cost of the project.

Of special interest to the landowners was the cost of construction, particularly since a private corporate venture was also under consideration. Reclamation officials assured them that the cost would approximate thirty-five dollars per acre, but certainly would not exceed forty dollars per acre. A number of issues were discussed, including the disposal of private land in excess of 160 acres, after which the landowners present were invited to step into an anteroom where abstract records of nearly every piece of privately owned land located within the proposed project area were spread out on tables. Notaries were present, and landowners were urged to sign stock subscriptions in the Yuma County Water Users Association, thereby placing a lien on each of their holdings for construction charges, a prerequisite before the government would begin work. It is estimated that 60 percent of the acreage was signed up on that day.[32]

Still, a sizable number of the farmers were undecided. Nevertheless, before the government would undertake the construction work, "practically all" of the private lands, in the opinion of the board of engineers, would have to be signed up. Some landowners were opposed to the necessity of disposing of their holdings in excess of 160 acres. Others were opposed because they held interests in the privately financed Ludy Canal. Some farmers expressed doubts about the ability of the land to bear the thirty-five-dollar-per-acre cost. Some regarded this figure as being rather high, particularly when there appeared to be a cheaper alternative.

The Colorado Delta Irrigation Company, organized by San Francisco capitalists and well-known hydraulic engineer James D. Schuyler, had already proposed to do what the government was suggesting, but at a substantially lower cost.[33] This proposed project, however, was confined to the Arizona side of the river, thus circumventing the complexities associated with the Indian reservation land (map 4.4). But the difficulties experienced by previous private irrigation enterprises in the valley, plus the fact that the current diversions from the Colorado were technically illegal, along with vigorous personal lobbying by those active in the Yuma County Water Users Association as well as by government engineers, persuaded the majority of farmers to accept the government's plan. By mid-November 1904, 56,332 acres out of a total of 60,632 acres, or 93 percent of the land in private holdings, were signed up.[34] Ultimately, 98 percent of all private land in the valley was brought under contract to take water according the provisions of the Reclamation Act. The remaining 2 percent was land mostly owned by Ludy and his associates, who opposed the project.

Linking the Imperial System

Events in the Imperial Valley paralleled those at Yuma, at least partially. In order to set the stage for the government takeover of the Imperial Canal system, William Smythe urged the people of the Imperial Valley to join the Yuma Project. To that end, the Imperial Valley Water Users Association was formed.[35] The Water Users Association was organized with the same objective as the Yuma County Water Users Association—to persuade farmers to support the government's plan. Problems similar to those experienced by irrigators in the Yuma Valley were beginning to befall Imperial Valley irrigators; that is, the ability to supply water to the mutual irrigation companies was becoming more difficult because no adequate filtering system with settling basins had been incorporated into the Imperial system, and the main canal began to fill with silt near the intake, seriously reducing the volume of water carried.

The problem had become so critical by the summer of 1904 that the Imperial Valley Water Users Association pressed Heber and Rockwood to sell

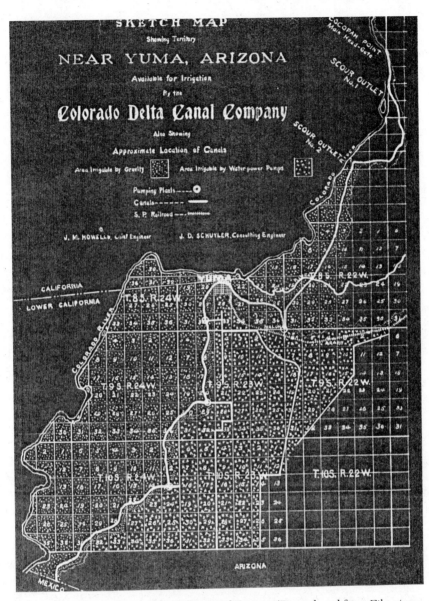

MAP 4.4. Proposed Colorado Delta Canal Project. (Reproduced from File 1/11, Box 16, Case 74, National Archives, Pacific Southwest Region, RG 21)

the California Development Company to the federal government so that the irrigation of the valley could be completed under the Reclamation Act.[36] Heber correctly accused the government of organizing numerous public meetings in the valley for the purpose of influencing its citizens to favor national irrigation and engendering negative feelings toward the California Development Company in order to drive it out of business. Heber eventually did acquiesce to the pressure and finally agreed to sell, but at a price of $5 million, whereas the association believed $1.5 million was a fair price. The lower the price the Water Users Association could negotiate, the better, for the purchase price and the cost of completing the system by the Reclamation Service would have to be borne by the landowners themselves. In the end, Heber and the association agreed on a compromise figure of $3 million, but the transfer would never be executed.[37]

Meanwhile, when the Reclamation Service challenged the legality of the California Development Company's Colorado River appropriation, Heber looked to Mexico for a new intake for the Imperial Canal. He was successful in obtaining a concession from Mexico, granting to the Mexican Company the right to divert, conduct, use, and sell water on both sides of the international boundary.[38] The agreement, ratified by the Mexican Congress on May 27, 1904, allowed the company to take ten thousand cubic feet per second out of the Colorado River on Mexican soil, but the corporation had to agree that Mexico would be allowed to use half the water flowing through the canal to irrigate Mexican lands in Baja California. The California Development Company then immediately dug a second heading just below the international boundary. This Mexican heading, like its predecessor in California, had a pronounced tendency to silt up. As soon as the summer flood of 1904 receded, it was discovered that deposition along the bottom of the canal had elevated it above the low-water stage of the river, once again threatening the water supply of the Imperial Valley.

Because of the excessive silting, which affected the upper four-mile section of the main canal, in October 1904 a second Mexican heading was completed four miles farther downriver, thereby cutting off the upper stretch of the main canal (map 4.5).[39] This was an expedient move because dredging the upper main canal would be a lengthy and expensive proposition, considering the volume of material that would have to be removed. Unlike the upper Mexican heading, the lower heading had sufficient fall from river to canal to give the water a scouring velocity, and the bed of this

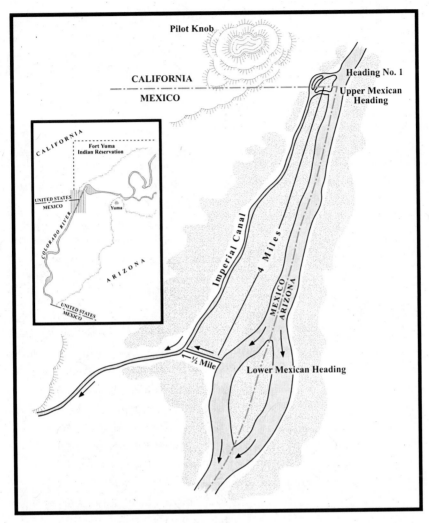

MAP 4.5. Colorado River headings of the Imperial Canal, spring 1905. (Based on C. E. Grunsky, "The Lower Colorado River and the Salton Basin," 25)

heading began to be enlarged by the water flowing through it. At first this condition seemed beneficial, since a large canal capacity was desirable in order to provide an adequate supply of water to the irrigators in the Imperial Valley. Hastily constructed, however, neither Mexican intake was protected by suitable headworks, although plans for headgates to control the flow at the lower Mexican intake had been drawn up and submitted to the Mexican government for approval.

Two months earlier, several Reclamation Service engineers were given the task of examining the Imperial Canal system and preparing estimates of its existing value, as well as the cost of installing permanent structures and completing the system, and determining the amount of irrigable land available in the Imperial Valley. The purpose of these investigations was to comply with the congressional resolution ordering an investigation of the valley, but also to determine whether the general conditions of the valley would justify absorption by the Reclamation Service.[40] After completing their inspection, the government engineers were unanimous in their belief that no diversion from the Colorado for irrigation could be permanently successful if provisions were not made for preventing the heavy silt from entering the canals, because under such conditions it would take an inordinate amount of dredging to keep the canals leading from the river open.[41] This, in fact, is exactly what was happening in the upper reaches of the Imperial Canal.

A Reclamation Service map was subsequently prepared illustrating the manner in which the Imperial Valley canal system could potentially be integrated into the Yuma Project via a main canal heading at the silt-removal works at Laguna Dam and constructed entirely on American soil, parallel to the international boundary (map 4.6). But this would require cutting a twelve-mile conduit beneath the sand hills located west of the river, a task whose estimated cost was $10 million, a prohibitive figure, it was thought.[42] Two months later, in October, the investigating committee of engineers commissioned by Congress returned to the Imperial Valley, this time accompanied by Senator Bard. While the inspection team was in the valley, the purchase of the canal system and its completion under the Reclamation Act were strongly urged by many of the water users.[43] One contemporary resident of the valley admitted that a "large majority" of the valley's settlers in 1904 wanted to turn the system over to the Reclamation Service.[44]

By now, however, the government's mood had changed. Despite its prior covetous designs on the Imperial Canal system, the Reclamation Service recommended against its purchase. Charles Walcott, director of the U.S. Geological Survey, the agency that oversaw the Reclamation Service, had already reported to the secretary of the interior that "from present knowledge of the conditions, a recommendation to pay three million dollars for the property and rights involved is not justified." Then, early in 1905, it was announced by the secretary of the interior that the assistant attorney general

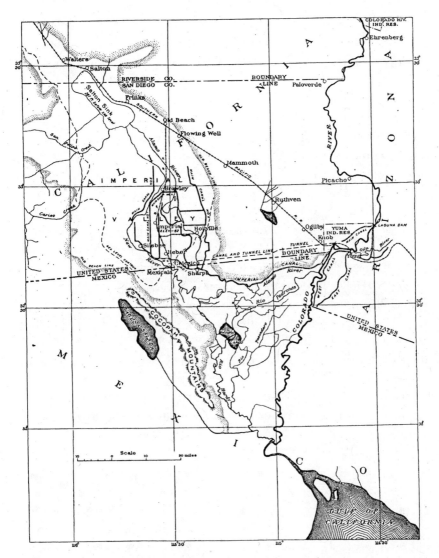

MAP 4.6. Proposed integration of the Imperial Valley into the Yuma Project via an "All-American Canal," shown by the dashed line located just north of and parallel to the international boundary. (Reproduced from Department of the Interior, U.S. Geological Survey, *Fourth Annual Report of the Reclamation Service, 1904–5* plate 9)

had concluded that no laws existed that could deal with the problem of carrying water through Mexico.[45] On the one hand, a canal located entirely on
American soil was not economically feasible, whereas, on the other hand,
one crossing Mexican territory was not politically feasible. The cost of taking
over the Imperial Canal system, combined with its international features,
therefore, precluded U.S. involvement in the Imperial Valley.

But some settlers already had begun to have second thoughts about sur-
rendering their system to the government anyway.[46] This reversal of community opinion began to develop after Heber made public his Mexican
agreement to Imperial Valley residents in a mass meeting at the end of
July 1904. Heber admonished the farmers for not supporting the company
when its title to Colorado River water had been challenged by the government: "You should have stood by the company . . . [when] your rights were
assailed, [instead] I was compelled to protect and enforce them single
handed."[47] Heber pledged to the crowd his determination to contest the
navigability of the Colorado in both Congress and the courts if necessary.
He told residents of the valley that he was completely confident that the
company would reestablish proprietary rights in the waters of the Colorado
River on the American side, for no court in the land would uphold this "fic-
tion of navigability."

The harassment by the Reclamation Service of the California Development Company, forcing it to seek protection of its water rights through negotiations with a foreign country, along with Heber's reassurance that work
was under way to alleviate the siltation problem, began to turn the tide of
public opinion against the government project. Late in 1904, after the Reclamation Service had decided to abandon its attempts to tie the Imperial irrigation system to the Yuma Project, support by Imperial Valley residents for
a government project collapsed. To the disappointment of many, it appeared
that the Yuma Project would be confined to the lands lying in the immediate
vicinity of Yuma, at least for the time being. But Heber's strategy to circumvent the government's new decree regarding water rights to the Colorado by
securing a concession from Mexico set the stage for an impending disaster
along the lower Colorado, one that would potentially imperil the completion
of the Yuma Project.

The Rampaging River

Heavy rains, which deluged the Southwest in the winter and spring of 1905,
resulted in an unprecedented sequence of floods in the Gila watershed.

These floodwaters washed into the Colorado River at Yuma, causing the Colorado below the Gila confluence to go on a rampage. The result was an erosion of the banks of the lower Mexican intake. Only three winter floods had ever been recorded since river gauging began at Yuma in the 1870s. But in the winter and spring months of 1905 alone, five heavy floods came roaring down the river. The first two floods, which occurred in February, caused the lower Mexican intake to silt up, requiring dredging to keep it open. In March the river reached flood stage again, but this time a widening of the intake occurred. As the season was approaching when the river's surface would normally be high enough to use the upper Mexican intake, prudence dictated that the lower heading be closed. Work on the dam was nearly complete when a fourth flood came charging downriver and swept it away. Another attempt was begun on the dam, but the fifth unseasonable flood washed away this work also.

It was now mid-April, and the normal flood season of the Colorado had not yet begun. By mid-June, as the summer flood was gaining momentum, the lower Mexican intake's 60-foot-wide original cut had expanded to 150 feet, and at least 11 percent of the river's flow was now running into the Imperial Canal.[48] Then as the floodwaters of the river receded, the banks of the intake began to cave in, as did the banks of the canal below the intake, and in August the Colorado broke entirely out of its channel and flowed unimpeded into the lowest portion of the Imperial Valley, the Salton Sink—soon to become known as the Salton Sea.[49] Portions of the Alamo River, and its distributary, the New River, the latter being formed where a breach in the bank of the Alamo River occurred ten miles above Sharp's Heading, as well as certain sections of the Imperial Canal, had now become the river's main channel (map 4.7). Because the flow of the Alamo was initially restricted by the structures located at Sharp's Heading, until the levees and diverting dams failed at this point it was the New River that was delivering most of the water into the Salton Sink.[50] The Colorado River, by adopting this new route, had taken the path of least resistance, choosing a 4-foot-per-mile grade to the bottom of Salton Sink, situated 278 feet below sea level, a much steeper route than the 1-foot-per-mile gradient that characterized its flow to the Gulf of California. And another heavy flood, occurring in late November 1905, had widened the intake to several hundred feet.[51]

At no time since river-gauge records had been kept at Yuma had the Colorado's discharge exceeded more than one-half of that recorded in 1905,

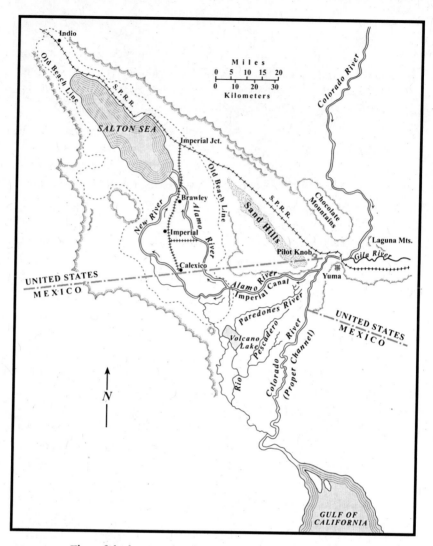

MAP 4.7. Flow of the lower Colorado River into Salton Sink via the Alamo River (Imperial Canal) and New River. (Based on C. E. Grunsky, "The Lower Colorado River and the Salton Basin," 15)

and its flow in 1906 proved to be equally prodigious.[52] In all, it took nearly two years and six different attempts to close the lower Mexican intake and to return the river to its former channel. After two unsuccessful attempts by the California Development Company to close the break in the spring of 1905, the Southern Pacific Railroad stepped in with financing and its own engineers to try to stem the flow. The Southern Pacific was impacted by the disaster in two ways: the rising water in the Salton Sea required continual changes in the location of the Southern Pacific tracks, an expensive proposition, and the Southern Pacific had served as the financial backer of the canal enterprises that were developing territory tributary to its railway system. An enormous loss of income would result if the tracks and Imperial Valley farmlands were inundated.

With its ability to lay track quickly and get the necessary construction materials to the site of the break, the railroad was in a good position to help. In late June 1905, at the request of the California Development Company, the Southern Pacific agreed to a loan of two hundred thousand dollars and to take over the management of the canal while it endeavored to close the crevasse. As soon as the summer flood of 1905 began to recede, work was started. It was decided to build a reinforced concrete and steel headgate at Pilot Knob, four miles above the break on the U.S. side of the border, which would ultimately become the permanent canal headgate for the Imperial system. The legality of diverting the Colorado by private corporations was no longer an issue; it had now taken a backseat to the necessity of regaining control of the river. At low water stage this headgate would permit the temporary diversion of the river into the main canal, which would have to be enlarged in order to carry the increased flow. This would leave the old river channel below, as well as the break, dry so that it could be closed. Once the break was closed, the concrete headgates could then be used to turn the river back into its proper channel and to control the amount of water to be diverted into the main canal.

The problem with this plan was that the summer flood season would come before the capacity of the main canal could be enlarged. Nevertheless, work on the concrete headgate was begun in mid-December. Rockwood, however, believed that a rapid rediversion to the river's channel was necessary because the irrigation system of the Imperial Valley could not withstand the severe strain for very many months.[53] His plan was to install a wooden headgate immediately beside the break capable of passing the low water flow of the river, divert the river through the headgate by constructing a barrier dam, then complete an earth fill dam to close the break. Rockwood's scheme was approved in mid-December, and work on both plans was carried out simultaneously. When Rockwood's gate failed in October 1906, his plan was abandoned. Meanwhile, a canal from the river to the new concrete headgate was excavated in September 1906, and the headgate was put into operation in early November (fig. 4.1). This allowed for the construction of a series of three rock dams to close off the break. Finally, by the end of 1906, all

FIG. 4.1. Concrete gate of the Colorado Development Company at Pilot Knob. (Reproduced from Sharlot M. Hall, "The Problem of the Colorado River," 330)

engineering works were in place and the river had temporarily been brought under control. Success was short-lived, however, for in late December another flood charged downriver. The three rock dams began to show evidence of weakness, and soon failed. The river was once again out of control.

A general feeling of panic now swept through the Imperial Valley, as farmers feared that if the Salton Sea, which had already grown to a 400-square-mile body of water, continued to be fed by the runaway Colorado, the entire valley would eventually become flooded and untillable.[54] The surface of the Salton Sea had climbed more than 70 feet in elevation in the course of one year since the river's rampage began. Another 90-foot rise would completely submerge the farming district around Brawley, located 110 feet below sea level near the south shore of the expanding inland sea. And the settlements at Imperial, El Centro, and Calexico, standing at successively higher elevations, would be next in line.

Although Imperial Valley farmland was not yet seriously threatened by the rising waters of the Salton Sea, significant damage had occurred to lands lying in the vicinity of the Alamo and New rivers, whose natural channels were not designed to carry the enormous flow that the Colorado was discharging into them. The threatened inundation of the main body of improved lands in the valley was prevented only by arduous work on dikes, which forced the water westward into the broad flat depression along the

New River, flooding lands in that district.[55] Valley residents now appealed to the Southern Pacific to ask the federal government to help control the Colorado. But if the government were to step in, treaty arrangements between the United States and Mexico would have to be drawn up, a process that would take too much time. President Roosevelt, therefore, appealed to the Southern Pacific to use all necessary resources to close the break, and verbally promised that the government would reimburse the railroad for the expense.

The Southern Pacific lived up to the challenge and finally closed the break with a more substantial set of dams in early February 1907. But in spite of Roosevelt's lobbying, legislation to reimburse the railroad was never approved because the reimbursement provision was included in a related bill that failed, a bill that revisited the possibility of the government's absorption of the Imperial Canal system. The bill would have appropriated two million dollars for that purpose. Part of that sum was to be used to reimburse the Southern Pacific Railroad. The remainder of the money would have been devoted to the establishment of an irrigation project in the Imperial Valley via an All-American Canal under the auspices of the Reclamation Service.

Ironically, Imperial Valley's congressional representative, presumably expressing the views of the majority of his constituents, argued persuasively for defeat of the bill. Valley residents had requested government assistance in river protection work only, not a government takeover of their irrigation system.[56] Even though thousands of dollars had to be spent annually by the California Development Company and the mutual water companies in the Imperial Valley for dredging to keep the canal systems in operation, landowners felt that the cost of Laguna Dam and the multimillion-dollar canal that would have to be constructed on the American side of the border was too high a price to pay for the government's project. Landowners also were fearful of the possibility of having to reduce their holdings to forty acres, the size of a reclamation project homestead. The incorporation of the Imperial irrigation system into the Yuma Reclamation Project seemed as elusive as controlling the Colorado River itself. Once again, the hope of welding the Imperial system to the federal project had slipped through the government's grasp.

In 1911, the California Development Company's remaining assets passed to the newly formed Imperial Irrigation District (IID). The Southern Pacific Railroad Company, the ultimate savior of the Imperial Valley from complete inundation, never was repaid for restoring the river to its former channel, finding itself on the losing end of valley residents' perpetual political feud with the federal government. Nevertheless, even without federal reimbursement, the railroad continued to benefit from the ongoing economic development of the valley, and the shipments that resulted therefrom.

5

Allotment

The contract for the construction of Laguna Dam was awarded in July 1905, and the work was to be finished within two years. Dissatisfied with repeated delays and increased costs of construction, the Reclamation Service took over the work in January 1907, when approximately one-third of the dam was complete. By July 1907, the original completion date, Laguna Dam was only about half-finished, and it wasn't until nearly two years later, March 1909, that the dam was put into service (figs. 5.1 and 5.2).[1]

Part of the delay was related to the errant Colorado during 1905 and 1906, the entire flow of which was emptying into the Salton Sea. This created a serious situation at Yuma because the immense energy associated with the spring flood of 1906 caused the river to cut a deep gully in the alluvial soil of the Imperial Valley, and the upper lip, or nick point, of the gully was migrating upstream at a rate averaging five miles per month, but in some places it was progressing a mile per day.[2] If the river were to remain unchecked and the nick point continued to migrate upstream, it was possible that at places below Yuma, where light alluvial material reached great depths, the downcutting could exceed one hundred feet, a depth that would make the river's diversion for irrigation totally impractical.[3] And since Laguna Dam rested on river silt on the bed of the river's channel, undercutting by the river would necessitate the abandonment of the Yuma Project altogether. Hence, active prosecution of work at Yuma was delayed until the river could be brought under control early in 1907.

In the fall of 1907, as Laguna Dam was taking shape, work began on the main canal heading on the Arizona side of the dam. According to the original plan, the Gila River would be crossed using an inverted siphon involving four reinforced concrete tubes laid three feet below the riverbed, through which water would be fed into the Yuma Valley. Within a week, however, this work was stopped. The recent floods on the Gila River made evident its tendency to change some part of its course when in flood, sometimes radically. And in extreme floods it would discharge more water and at higher

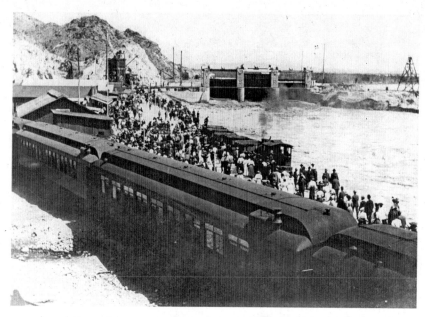

FIG. 5.1. View of sluiceway gates on the California side of Laguna Dam with arrival of a special train during the christening of the dam, March 31, 1909. (National Archives photo, courtesy of the Yuma Area Office)

FIG. 5.2. View of Laguna Dam headworks, California side, May 1910. (National Archives photo, courtesy of the Yuma Area Office)

FIG. 5.3. Nearly complete Yuma Project main canal passing beneath the Southern Pacific tracks. (Reproduced from Francis L. Sellew, U.S. Reclamation Service, "Yuma Project Historical Sketch, 1902–1912")

velocity than the Colorado. The Colorado River, by comparison, was much less erratic and tended to give some warning of an impending flood by rising more slowly. The Gila crossing was therefore abandoned because the difficulty of holding the Gila River banks at the ends of the underground siphon was considered too great; the main diversion works were therefore shifted to the California side of Laguna Dam, with the inverted siphon crossing the Colorado at Yuma (see map 1.2).[4]

By March 1910, a portion of the main canal covering the reservation division was ready to be placed into service (fig. 5.3). Meanwhile, in 1907 the Southern Pacific Railroad Company, after closing the Mexican break, laid track on top of the newly constructed levee from Yuma to the Potholes. The track was installed primarily to haul materials and supplies to and from Laguna Dam, but it was also built to serve the reservation division of the Yuma Project, an area anticipated to be under intensive cultivation soon.

But there was more to making the Yuma Project a productive agricultural region than overcoming the engineering challenges; before bringing the reservation division of the Yuma Project into cultivation, a number of human concerns had to be dealt with, the most prominent being Quechan allotment and the opening of the surplus reservation lands to non-Indian settlement. The first step was to conduct a census of the Quechans so that the process of allotment could get under way.

Quechan Census

In late June 1904, Lippincott wrote chief engineer Newell requesting that "a census be promptly made of these Yuma Indians for the purpose of making the allotment necessary subsequent to the constructions."[5] But since the construction of the irrigation works was moving forward more slowly than anticipated, apparently there was no sense of urgency to initiate the allotment process, and a year later little had been accomplished. Finally, in mid-July 1905, time was drawing near when it would be necessary to know where the Indian lands would be located because the irrigation laterals would have to be planned so as to harmonize with the system of land subdivision used to allot the Indians. The secretary of the interior was subsequently informed by the Reclamation Service and the U.S. Geological Survey that the allotments to the Indians should be made in a compact body, not only for reasons of economy of construction but also because "it would not be advisable to have Indian lands distributed among the lands of white settlers."[6] Although the purpose of the General Allotment Act was to advance the assimilation of Indian tribes into the white community, the segregation of the Quechans on the Yuma Project was the main concern of the Reclamation Service. It will become increasingly apparent that numerous government agencies exhibited little interest in the welfare of the Quechans, but plenty of interest in their land.

Meanwhile, a census of the Quechans, conducted by U.S. Indian agent John Spear, who also served as superintendent of the reservation, had been completed by the end of June 1905. Spear's count totaled 675 Quechans dwelling on the reservation, another 69 Indians living in the Yuma Valley near Somerton, and 82 Indians residing in Mexico, for a total of 826 Quechans who potentially could receive allotments.[7] Those living in the Yuma Valley came to be known as the Somerton homesteaders, for they had entered forty-acre parcels under the 1884 Indian Homestead Act. The first group totaling twelve families filed their entries in October 1901, and four more families filed entries in July 1903. Presumably, they had moved to Arizona to escape compliance with regulations to which they would be subject on the reservation.[8] The community quickly became a refuge for other Indians desiring to escape reservation regulations. In an effort to stem this flow, an order was issued in November 1905 to include these Indians under the supervision of the superintendent of the Fort Yuma Indian Training School.[9] Many of the Indians counted in Mexico were working there to help close the Mexican crevasse of the California Development Company, but some lived there also to avoid sending their children to the Indian school on the reservation.

Lippincott, when he learned of the results of the census, was thoroughly displeased with the count. It was his original understanding in 1904 that there were approximately 711 "Yuma Indians." After the count he wrote Newell requesting a copy of the census, stating that there was a greater number of Indians than expected, and in a callous tone declared that he wished "to avoid the allotment of high priced lands to Indians who are really not entitled to it."[10] Lippincott had apparently lost sight of the fact that first and foremost this was Quechan land; certainly, an accurate count was required, but careful measures had to be employed to guarantee that those who were entitled to allotments would get them.

The secretary of the interior followed up Lippincott's concerns with a letter to C. F. Larrabee, acting commissioner of Indian affairs, requesting answers to a number of questions, including the status of the 82 Indians living in Mexico, whether the Somerton homesteaders were entitled to allotments, and exactly what was the correct figure—711 or 826? The acting commissioner was vague in his response. He confirmed that the census showed 826 Indians, but when it came time for allotments to be made, he said that "there may be more or less than that number." If it was determined that the 82 Indians residing in Mexico belonged on the reservation, then they were entitled to allotments. And if the Somerton homesteaders were deprived of certain benefits because of the construction of the Yuma Project, then they also should be given allotments. Larrabee continued by stating, "There is no way whereby the number can be fixed in advance, so that it will not vary, as naturally births and deaths will occur, and the Indians being of a roving nature, some not listed may be found when the allotting agent goes to work."[11]

The Somerton homesteaders particularly troubled the Reclamation Service. Their situation had not been accounted for in the course of planning the Yuma Project. John Spear, in a letter to the commissioner of Indian affairs, predicted that the land on which they had filed would soon be very valuable, "and if the Indians can only hold it a few years and get water, until they learn to farm by irrigation, they will be all right."[12] But according to the Indian homestead law, such homesteads would be held by the United States in trust for the Indian entrymen and their heirs for a period of twenty-five years. These lands therefore could not be subjected to the annual construction charges required under the Reclamation Act, for that would be viewed as a lien against these properties. The Indians legitimately would have to be exempted from such charges. The Yuma County Water Users Association, however, had already informed the secretary of the interior that the existence of the Somerton homesteaders should not increase the cost of reclamation to private owners. Yuma Valley water users felt that the government should provide for the cost of irrigating the Indian homesteads since it retained title to the land being held in trust and was, in this sense, acting as the guardian to these Indians.[13]

Another potential problem involving the Somerton homesteaders focused on the construction of the Yuma Valley protection levee. Unlike non-Indian farmers in the valley who perceived the Colorado's annual overflow to be a menace, the Somerton homesteaders depended on the summer flood for their agricultural pursuits, although most did not rely on agriculture for a living. Their principal livelihood was derived from hiring themselves out as day laborers on neighboring ranches. The consensus among Reclamation officials nevertheless was that the Somerton homesteaders' economic status would improve once their farms were served with water, and that they therefore should not be entitled to allotments on the reservation. On the other hand, district engineer Homer Hamlin suspected that allowing the Somerton homesteaders to remain in the heart of the Yuma Valley was probably not a wise move either and would "greatly displease the adjacent land owners, as certainly no one will wish to own property adjoining such a tract."[14]

Dissatisfied with the results of Spear's census, the Reclamation Service conducted its own count. Surprisingly, the latter total turned out to be higher than the former: 840 versus 826 Quechans.[15] The difference was accounted for by the Indians living in Mexico. The Reclamation Service identified 96 Quechans living in Mexico compared to the Indian Service's 82. Nevertheless, Lippincott, apparently concerned about cost overruns on the project, felt that only the 675 Quechans who were actually living on the reservation should be allotted: "If the Indians off the reservation are not given an allotment it will make a difference of $30,000 to us," he wrote to chief engineer Newell early in 1906.[16]

Lippincott's attitude, and that of the Reclamation Service generally, was not inconsistent with that of the administrators of the allotment process. William Jones, who served as commissioner of Indian affairs from 1897 to 1905, responded to the growing demand for white access to Native property by supporting an acceleration in the pace of allotment and working to open

the tribal domain through land cessions and changes in Indian Office procedures.[17] During his term in office, he pledged that "the pressure for land
must diminish the reservations to areas within which [the Indian] can utilize
the acres allotted to him, so that the balance may become homes for white
farmers who require them."[18] Hence, both the Reclamation Service and the
Indian Office worked to keep allotments as small as possible so that surplus
lands could be distributed to whites.[19]

The commissioner's actions were facilitated by the Supreme Court ruling in 1903 in *Lone Wolf v. Hitchcock* that had given Congress the power
to take Indian lands without their consent.[20] Originally, allotment was to be
initiated at presidential discretion, and the tribes had to consent to the sale
of their surplus lands before the allotment process would be carried out.
Since Congress now had authority over allotment, it summarily neglected to
inform the Quechans of the impending legislation or to seek their approval
for the disposal of surplus lands when that body sanctioned the allotment
of the Fort Yuma Indian Reservation in April 1904. But this failure to com-
municate might also in part have been due to the widely held view that the
Quechans had already assented, in the 1893 agreement as ratified by the
1894 act, to five-acre allotments, even though the Colorado River Irrigation
Company that was to deliver water to the reservation had long since failed.
Presumably, Congress tacitly assumed that the 1902 Reclamation Act, in
conjunction with the act of April 21, 1904, which approved five-acre allotments for the Quechans, allowed the government to fulfill the commitment
to furnish water on the reservation that originally was to be supplied by the
private corporation. Either way, Congress apparently felt no obligation to
consult with the Quechans about the matter.

Location of Quechan Allotments

While the exact number of Quechans entitled to allotments was being determined, that portion of the reservation to be allotted also had to be established. Article II of the 1894 agreement stipulated that the Quechans
would be entitled to select and locate the lands that would be allotted, and
Article III provided for the disposal of surplus lands after the allotments
were approved. In April 1907, Levi Chubbuck, working for the Indian Service, informed the secretary of the interior that the location of the Indian
allotments should be determined without delay. He pointed out that since
the Indian allotments were only five acres in size compared to what probably would be forty acres for whites, it was imperative that the best land

on the reservation be retained for the Quechans and that the allotments be made in such a way that the Indians be grouped together. Inspector Chubbuck also recommended that the strip of woodland located outside the levee adjacent to the river be attached to the Fort Yuma School as a timber reserve. At the same time, however, Chubbuck was pessimistic about the future of Quechan irrigation agriculture: "The Primitive Type of Agriculture practiced by the Yuma Indians—consisting in sowing of seed on the sun-cracked surface of the bottom lands after the recession of the floods," wrote Chubbuck, "will make it difficult to induce these people to quickly adopt methods that must be used on 5-acre allotments under irrigation. In fact it need not be expected that much will be done by the allottee with their land for a number of years."[21]

Meanwhile, Ira Deaver, who had succeeded John Spear as superintendent of Fort Yuma Indian School, informed the commissioner of Indian affairs that the Indians were being uncooperative regarding the portion of the reservation on which they wished to live because one faction of the tribe objected to being allotted. However, after consulting with "a great number of the more progressive Indians," the location was determined that would "be most acceptable to all those who will accept the allotment willingly."[22] The tract so designated was situated adjacent to and north of the Southern Pacific Railroad right-of-way, a parcel containing some of the most level and fertile land on the reservation. Deaver recommended that allotment be done at once because it would take considerable time and labor to prepare the land for irrigation. He also recommended, as Chubbuck had done, that the strip of land located between the reclamation levee and the river be reserved for the Indians as a wood and timber supply. Finally, he urged the establishment of a townsite on the reservation so that the proceeds from the sale of lots could be used to benefit the Indians.

In June 1907, W. H. Code, now chief engineer of the Indian Service, who five years earlier had been charged with devising an irrigation plan for the reservation, was now assigned the task, in collaboration with the new supervising engineer in California, Louis C. Hill, and superintendent Deaver, of making a tentative selection of a body of irrigable land for Indian allotments. Code pointed out that because he had conducted surveys of the reservation in the past, he and Deaver were equally familiar with the best lands under the proposed canal and recommended that selections be confined to T16S,R22E, which contained approximately "6,000 acres of the choicest land on the Reservation."[23] He advised that a capable allotting agent should be

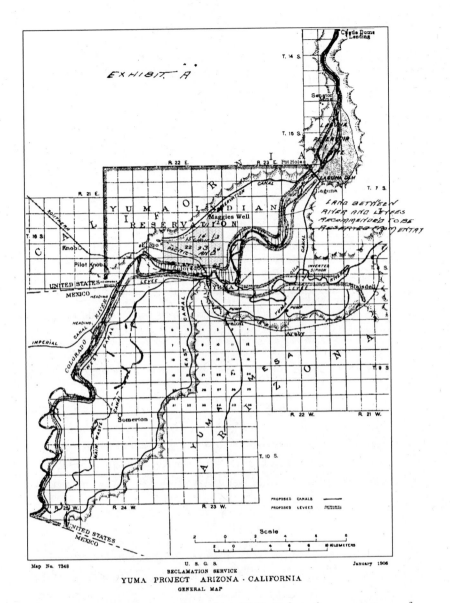

MAP 5.1. Map accompanying letter from W. H. Code to the commissioner of Indian affairs showing lands *(highlighted)* recommended for Quechan allotments. (Reproduced from File 154-A, Box 1058, National Archives, Washington, D.C., RG 115)

assigned to perform the difficult task of allotting the Quechans, and that his work should focus on sections 11–15 and 22–28 of the designated township (map 5.1). The lands in this portion of the reservation were indeed among the best from the standpoint of ease of leveling and their perceived productivity according to the government soil survey published in 1905, as well as their proximity to Yuma, where many of the Quechans worked odd jobs. In terms of the parcel of land identified for allotment, Code, Deaver, and the commissioner of Indian affairs were unanimous in keeping the best interests of the Quechans in mind.[24]

In September, special allotting agent Douglas Graham was appointed for the purpose of surveying the land and proceeding with the allotments. Graham's job, although presumed to be a difficult one, turned out to be much more demanding than anticipated. Agent Graham subsequently spent several months during 1907 and 1908 on the reservation, preparing family history cards and making allotment corners of the five-acre tracts. But in December 1908, the Quechans began to have second thoughts about the allotment process, protesting that they were unwilling to accept allotments in severalty until they saw with their own eyes the water turned from the Colorado River into the irrigation ditches. Out of frustration Graham left the reservation, but he was called back in July 1909. The water was now running in the canals, and he was informed that the majority of Quechans were finally willing to accept their allotments in severalty. Not long after he returned, however, Graham found there was now a move to seek legislation to increase the size of the Quechan allotments from five to ten acres, and again the Quechans were resisting allotment, this time pending the outcome of the campaign to double the size of their individual holdings.[25]

Campaign for Ten-Acre Allotments

It was the Indian Rights Association, based in Washington, D.C., that initiated the crusade to push for ten-acre allotments. Similar to other organizations of concerned white citizens who lobbied for what they viewed as the interests of the Indians, the Indian Rights Association favored Indian assimilation. Its mission was not to protect traditional Indian life but to promote government aid in helping Indians adopt the Anglo lifestyle. The organization, formed in 1882, was interested in protecting Indian resources only to the extent that these resources were necessary to the process of assimilation.[26] Samuel Brosius, the association's Washington lobbyist, had apparently received numerous letters from the Quechans concerning their lands and their allotments. In late August he visited the reservation and held a council with the Indians, after which he wrote Robert Valentine, commissioner of Indian affairs, urging his support and the adoption of a plan "to allot not less than ten acres of irrigated lands to each member of the Yuma tribe of Indians."[27] Brosius emphasized that the existing law affecting the Quechans was enacted without consulting the tribe, and, hence, the Indians had not consented to the allotments.

In an attempt to garner public support for the cause, Brosius had written a letter to the editor of the *California Independent,* a copy of which he included with the commissioner's letter. "By no other law, so far as recalled, has Congress directed so MEAGRE [*sic*] AN ALLOTMENT to Indians, especially where they own their lands," he wrote.[28] He conceded that Congress had the power to allot the reservations and sell off the surplus lands, but "as guardian of the Indians, the increased responsibility is assumed thereby to see that these wards of the nation are fully protected in every way."[29] Brosius advised the commissioner that his sources in the Yuma Valley maintained that no one could live comfortably on five acres of irrigated land. Even though the average Indian family might comprise three, four, or five individuals, twenty to twenty-five acres still would not be enough land, particularly in light of the forty-acre tracts that the Reclamation Service had recently approved as an adequate-size family unit in the area to be opened to non-Indian farmers.

Brosius argued that because the Quechans were widely recognized as being among the most backward of tribes, instead of receiving less acreage, they should be allotted more than white settlers, who presumably were more highly skilled farmers. He insisted that the five-acre allotments were too small to enable the Quechans to progress toward civilization. The only type of farming that could possibly succeed under such circumstances was truck farming, and to expect the Indians to turn to this type of farming was simply impracticable considering the level of their farming technology, the lack of a local market for the produce, and the high freight charges involved in shipping the products to distant markets. "From every standpoint," wrote Brosius, "it seems grossly inequitable that the Yuma Indians—THE PRESENT OWNERS of the land—should be compelled to accept a FIVE-ACRE ALLOTMENT."[30] He then apprised the commissioner that the Yumas strongly objected to this "threatened wrong" and hoped that their appeal to justice in order to secure not less than ten-acre allotments would be honored. Finally, Brosius assured the commissioner that if larger allotments were approved, there would be no need to delay the opening of the Yuma Project since sufficient land could be reserved for the Indians in the designated area of the reservation while still leaving an ample amount of land to be sold to non-Indian settlers under the terms of the Reclamation Act.

Brosius's original appeal to the commissioner of Indian affairs was supported by a number of other pleas, largely by religious organizations, to increase the size of allotments. The Methodist missionary stationed at the Yuma Indian Reservation related that he was "heartily in sympathy" with Brosius's work on behalf of the Yuma Indians, and was glad that something was being done to bring their needs before the public. The Catholic bishop of Los Angeles also urged Valentine to support larger allotments. And at the Los Angeles meeting of the Woman's Missionary Society, a motion was adopted advocating ten-acre allotments for the Quechans.[31]

Meanwhile, in mid-September Capt. Simon Miguel, chief of the Quechan tribe, who represented the faction opposing allotment, drafted a letter, most probably with the assistance of the Indian Rights Association, apprising Valentine of the Quechans' view of allotment. Miguel opened his letter to the commissioner by stating that he was writing on behalf of his people to learn whether the decision had definitely been made to allot a portion of the reservation, and if so, whether the size of the allotments had already been determined. Miguel wrote: "If there has been no final decision for the opening of our reserve, will you not please hold the matter open so that you may hear our arguments and have time to look into our case. Our tribe, almost to a man, opposes the opening of the reservation. It cannot see how the government can take from its members, without their consent, the land that it told them should be theirs forever."[32]

Miguel noted that he was present in 1884 when the reservation was originally granted in perpetual trust to the tribe, and that this was the only outstanding compact to which the tribe had consented. He then summarized the arguments, espoused by Brosius and others, regarding the reasons that five-acre allotments would be too small. Miguel vowed that the tribe would do anything in its power to prevent the five-acre allotments, but that if allotment could not be prevented, "we wish the allotment unit large enough to enable us to hold the entire reservation." This latter request, of course, ignored what was now the primary objective of the allotment process—to open surplus Indian land to non-Indian homesteading. Miguel beseeched the commissioner to at least extend the courtesy of informing the tribe of where it stood regarding allotment. He concluded: "Our reservation is not very large and the loss of any part of it makes our holdings entirely too small. We have been very unfortunate in never having been able to bring our wishes before the Indian Department. Our agents have never done much for us in that line, probably were afraid to do so. At present we know nothing, we have heard much, but what we have heard is not good."[33]

Kansas senator Charles Curtis, who served on the Committee on Indian Depredations, also wrote Valentine, expressing his conviction that five-acre allotments were too small, that the Quechans should have at least ten acres each, and that there would still be plenty of land left over for sale.[34] He requested that the commissioner delay making allotments until something could be done by the friends of the Indians at the next session of Congress, for it was his understanding that the California delegation would offer an amendment to the existing law. In early October, Valentine informed Newell of the recent developments. Newell in reply said that he would look over the matter with care and comment in detail at a later date. In the meantime, he requested information on the total number of individuals to be allotted, the probable size of the Quechan family, and the probable number of families or groups living together. He also wished to be informed about the status of the allotment process and the total amount of land being farmed by the Indians.

Before the requested information arrived on his desk, however, Newell apparently had already formed an opinion regarding the size of allotments, couching part of his argument in monetary terms. In a seven-page letter, Newell pointed out to the commissioner that in light of the April 21, 1904, statute providing for five-acre allotments, ditches were already under construction to irrigate approximately sixty-nine hundred acres, the amount of land that had been reserved for the Indians.[35] Newell rejected the argument being circulated that the Indian should have more land because he is an inferior farmer, and suggested that this was analogous to the argument that an Indian, who was unable to handle one team of horses successfully, should have two teams of horses, incurring the extra expense in feeding and care of two teams because he cannot handle one.

In any case, Newell pointed out that once the land was placed under irrigation, operation and maintenance charges would be incurred at one dollar per acre per year. By doubling the size of allotment, these charges would also double. Reclamation Service information, according to Newell, suggested that Quechan family size ranged generally between four and six, meaning that they would receive between twenty and thirty acres per family. This was equal to or more land than the twenty acres that the agricultural experts in the region claimed was sufficient to support the average non-Indian family. The expense of simply maintaining ditches would tend to keep parcels small, Newell advised.

Newell then expressed his own view of what would become of the Indian allotments:

> The object of allotting to these Indians five acres each was to give them an area which might possibly be utilized, altho [sic] not probably needed. Very few of the five-acre allotments will be cultivated for many years, and if the land is put to use, it is presumable that it must be through some form of lease to white men. The only object of giving more than 6,900 acres of irrigable lands to the Indians would apparently be to increase the area which can be leased, or to permit the Indians to speculate on the future increase in value of the land.

Newell did not mean to infer that this situation would result from Quechan laziness; on the contrary, because of their reputation as hard workers, Indian labor was in considerable demand, and most Quechans would prefer to continue to hire themselves out as day laborers in lieu of adopting irrigation farming. Some also would no doubt prefer to continue to raise crops as in the past, planting on the overflow lands outside the levees. It was Newell's opinion, therefore, that "it would be a great mistake to attempt to change existing law, and I doubt whether on full presentation of the facts that Congress will act in their discretion."[36]

F. H. Abbott, acting commissioner of Indian affairs, had succeeded Valentine during the month of October and was now handling the allotment question. Even though Newell had already reached his own conclusion regarding the size of Quechan allotments, Abbott transmitted to the Reclamation Service the information that Newell had originally requested. Abbott confirmed that the allotments had not been completed, largely because the Indians "in a great many cases" had refused to give the allotting agent the necessary information regarding their family histories.[37] But based on information on file in his office, there appeared to be approximately six hundred Indians living on the reservation entitled to allotments, and Quechan families, as Newell correctly suggested, generally ranged in size from three to five members each. Abbott was not able to ascertain the amount of land that the Quechans had under cultivation.

The following day, Abbott informed the chief engineer that he had received communications from Senator Frank Flint from California and others that there would be a movement during the next session of Congress for ten-acre allotments. "In view of the requests for the suspension of allotment work on this reservation until Congress can consider the matter further," wrote Abbott, "and in view of the high standing of the persons from whom some of these communications originate, I feel compelled to suggest that, out of interest of all parties concerned, it would be advisable to take no further steps with regard to allotments to the Yuma Indians, under existing law." The acting commissioner then addressed Newell's belief that upon full presentation of the facts Congress would not enact additional legislation: "This may be true, but I feel that at least an opportunity should be given that body to consider the matter further and enact such additional legislation, if any, which it may desire."[38]

Newell, however, was surprised to learn that there were only about six hundred Indians entitled to allotments, for his own census counted significantly more potential allottees. Newell informed Abbott that the assumption had been that there were many more than six hundred; consequently, sixty-nine hundred acres had been outlined on Reclamation Service maps as available for Indian allotments. Abbott, having just taken over as acting commissioner, perhaps was not fully informed about this particular case, for he must have believed that only those Indians currently living on the reservation would be given allotments. Newell then pointed out that if there were six hundred Indians only, each entitled to ten-acre allotments, sufficient land had already been reserved, whether the allotments turned out to be five or ten acres in size. "It does not seem necessary, therefore," Newell suggested, "for us to delay taking up the remainder of the land, if over 6,000 acres may be held temporarily pending action by Congress on the matter."[39]

Although on the surface it appears that Newell had acquiesced to the ten-acre allotments, he undoubtedly knew that the six hundred figure was far too small a number. Neither he nor Louis Hill, who, in July 1906, had succeeded Lippincott as Reclamation supervising engineer in California when Lippincott left the service to become chief engineer on the Los Angeles aqueduct project, would willingly give up on the original five-acre allotments. Hill noted in a letter to Newell, "We have forwarded you, at various times, letters stating somewhat the habits of the Indians on the Yuma Reservation," and that careful inspection of the reservation showed evidence of less than two hundred acres total having been under cultivation, making a very small showing per individual. Hill continued:

As an average I would consider that five acres per individual for a tribe of Indians such as the Yuma Indians are, and located in a country where the soil is so extremely fertile and the growing season is long, was too much rather than too little, and any attempt to enlarge this area will not further the interest of the Indians at all. From the point of view of any outsider looking for the good of the Indians themselves this is about the worst condition that could possibly exist. I do not believe in making the size of holdings for Indian families larger than can well be handled by the Indians with their capital, ability and willingness to work. I think it would be far better for them to have less land than they would like to have rather than more than they would need.[40]

Where engineer Hill acquired his presumed expertise to expound on the best interest of the Quechans is unknown, but his condescending attitude toward the Indians and the fate of their reservation appeared to make him a worthy successor to Lippincott. The Reclamation Service had no business being involved in this debate anyway; the size of the allotments was really a matter between Congress, the Quechans and their representatives, and the Office of Indian Affairs. The underlying bias of the Reclamation Service derived from the economic feasibility of the project, and this was largely based on how much surplus reservation land would be available for settlement by non-Indians.

Newell responded by saying that he was glad to have the information contained in Hill's letter. He noted cynically that there would be an attempt made to double the area allotted to the Indians "simply on the ground that the so-called friends of the Indians wish to do twice as much as the Government recommends. If we had set the acreage at 10 acres they would probably have urged 20, and if we had suggested 20 they would have clamored for 40," wrote Newell. He then instructed Hill to compose a letter that he could read before the Senate Committee on Indian Affairs containing arguments concerning parcel size that, from the standpoint of the future of the Indians, would be most to their advantage. Hill, in his reply, repeated his view that smaller was better:

There will be a large number of Indians who will never be farmers. If the allotment were made on the basis as specified by law there is no question in my mind but that there would be a large amount of land left vacant for a great number of years, because many of these Indians can earn good wages for as many days each month as they care to work, and they much prefer to do this work to farming. Farming has not been their natural way of obtaining a livelihood.

From my observation . . . I would say that it will be far better to have a small acreage than a large one. It requires capital in addition to labor to make a success of farming today, especially in an irrigated district. Hard work alone will not enable a man to handle large tracts of land.

Newell, however, apparently was uncomfortable with this line of reasoning. Perhaps he felt that this argument was not consistent with the original intent of the General Allotment Act—to turn Indians into farmers—so instead he resorted to raw economics in his effort to persuade the commissioner of Indian affairs to reject ten-acre allotments. In February 1910, he cautioned the commissioner that if the proposed change was made, it would materially modify the plans that had already been completed by the Reclamation Service, thereby requiring additional funds to irrigate the increased area set aside for the Indians. Newell then detailed for the commissioner the additional costs, which totaled $178,750, that would have to be carried by non-Indian settlers in order to repay the cost of construction.[41]

But the issue before Congress was not about the plight of non-Indian settlers; rather, it was about fairness to the Quechans, at least within the constraints of both the Newlands Reclamation Act and the General Allotment Act. In the end, the friends of the Indians presented a convincing argument, and the Committee on Indian Affairs, citing a letter from the secretary of the interior recommending the ten-acre allotments, reported favorably on the Senate bill that amended the act of April 21, 1904, to increase allotments to ten acres. But by the time the proposed increase in allotments was given final approval, the annual Indian appropriation bill, to which this amendment would be attached, had already been ratified. Hence, nearly one year elapsed before the following year's appropriation bill was presented for approval, and on March 3, 1911, Congress ratified the increase in allotments from five to ten acres.[42] A small sum of money was also included in the amendment to cover the first year's increase in irrigation costs resulting from the increased size of allotments.

Closure

Eight months later special allotting agent Charles Roblin, who had succeeded Douglas Graham, was dispatched to the reservation to survey and dispose of the ten-acre allotments to the Quechans. His work began in early January 1912, and before the end of April he had completed the allotment process. The allotments totaled 8,190 acres and were located on the irrigable land north of the newly constructed levee in T16S,R22E, the parcel recommended by Code and Deaver.[43] Family allotments, with few exceptions, were arranged in contiguous blocks (map 5.2). The ten-acre Indian allotments translated into an average of just less than thirty acres per family, smaller than the reclamation homesteads taken up next door in T16S, R23E. But parcel size by family unit was quite variable, ranging from 101 ten-acre parcels to several that were more than one hundred acres in size (table 5.1).

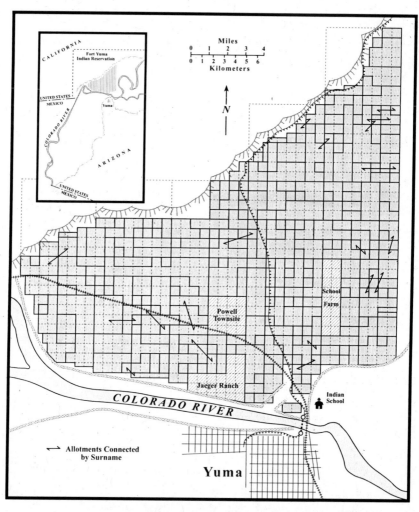

MAP 5.2. Distribution of Quechan allotments by surname. (Compiled from "Map of Yuma Indian Reservation: Allotted Lands," Irrigation and Water Rights Case Files, 1911–1957, National Archives, Pacific Southwest Region, RG 75)

TABLE 5.1
Frequency of Quechan family units by parcel size

PARCEL SIZE (ACRES)	NUMBER OF FAMILY UNITS
10	101
20	56
30	32
40	26
50	17
60	12
70	7
80	4
90	4
100	2
110	3
120	1
130	1
140	0
150	0
160	0
170	0
180	0
190	0
200	1

SOURCE: Compiled from "Map of Yuma Indian Reservation: Allotted Lands," Irrigation and Water Rights Case Files, 1911–1957, National Archives, Pacific Southwest Region, RG 75.

The largest family unit encompassed two hundred acres. In reality, the irrigable acreage in many cases was less than the ten acres allotted each individual, since the irrigation works themselves consumed considerable land area. Originally, several sections in the western- and southernmost part of this township had been surveyed for the purpose of opening them to Anglo settlers in farm units of forty acres each, but this survey was abandoned after it was determined that additional land would be needed for the Indians. Nearly a year and a half after Roblin's work was completed, the secretary of the interior approved the Indian allotments, and in early February 1914, Indian trust patents were issued. This left approximately 7,500 acres of surplus irrigable land available to non-Indian homesteaders.

Fortunately for the Quechans, the allotted land has retained its "inalienable" status, effectively keeping it from passing into the hands of non-Indians. Nevertheless, the size of their reservation shrank almost 80 percent due to the disposal of surplus lands; transfer of unsold, unallotted lands to the Bureau of Land Management; and flood control and irrigation projects.[44] And, as predicted by numerous contemporary observers close to the scene, many of the Quechans would shun irrigation agriculture, and their allotments would come under lease to non-Indian farmers. The promotion of Indian agriculture and assimilation into the American mainstream had long since been abandoned by the government as the primary objectives of the Allotment Act; instead, the obvious intent of the law had shifted to one facilitating the transfer of arable Indian lands to white settlers. Congress, the Department of the Interior, and the Office of Indian Affairs were all of like minds in this regard.

Hence, potentially productive irrigable areas found on the reservations in the arid West represented an important component of the final frontier. The official view of Native property, according to Frederick E. Hoxie, had switched from the idea that it was a birthright toward the notion that it represented a part of the public domain, and, as such, Indian lands should foster regional economic growth rather than serve the narrow needs of their "backward" inhabitants.[45] By doubling the size of the original Quechan allotments, Congress and other government agencies associated with the allotment process could perhaps absolve themselves of some of the underlying guilt that might have subliminally been attached to their actions—in this case, a further reduction in the territory encompassed by Quechan land so that reclamation homesteads could be carved out of the "surplus" reservation land by non-Indian farmers.

6

Bard

The irrigable land in the Fort Yuma Indian Reservation overlapped portions of two townships. Since all of the Indian allotments were confined to that portion of the Colorado River floodplain located in T16S, R22E, Anglo settlers by default were assigned the irrigable area found in the adjoining township, T16S, R23E. The former came to be known as the Indian unit of the reservation division, whereas the latter was appropriately labeled the Bard unit, after the senator whose untiring efforts made the Yuma Project and allotment of the reservation a reality. It was actually the new townsite established by the Reclamation Service to serve the non-Indian unit that originally was called Bard. In January 1909, director Newell suggested that the new townsite be named in honor of the then former senator, not only because of the pivotal role he played in gaining approval of the Yuma Project but also because the name was "short and distinctive."[1] In time the entire unit devoted to non-Indian settlement took on the Bard appellation.

The Bard unit represented that part of the project for which the Reclamation Service held especially high hopes, primarily because this portion of the Yuma Project was not encumbered by complications involving the Indian allotments located just to the west, or by settlers and their canal enterprises that were found across the river in the Yuma Valley. The Bard unit encompassed vacant land; hence, it was the Reclamation Service that would lay the foundation upon which the process of settlement and development would unfold. But the settlement policy of the service turned out to be extraordinarily shortsighted. Once the engineering works were in place and a systematic plan for land disposal was formulated, the Reclamation Service assumed that all other facets of colonization would simply fall into place. Unfortunately, this sanguine scenario was not to be the case.

The Farm Unit

The first step in the process of land disposal was to ascertain the optimal size of the farm unit within the context of the environmental conditions under which the Yuma Project was being developed. It was specified under the original Reclamation Act that reclamation homesteads would be 40 acres in size, but the original act was amended in June 1906, leaving the size of the farm units within reclamation projects to the discretion of the secretary of the interior. Hence, there was considerable debate over the issue of farm unit size at Yuma before the final decision was made, a dialogue, one might add, that was absent when the original decision regarding 5-acre Indian allotments was made. In the fall of 1907, as work on the main canal heading on the California side was about to get under way, both Newell and A. P. Davis, second in command at the Reclamation Service, sent requests to a number of individuals, including supervising engineer Hill, and R. H. Forbes, director of the agricultural experiment station of the University of Arizona, seeking their advice regarding what they believed the size of the farm unit should be. After a review of their responses, Newell and Davis would then forward their recommendation to the secretary of the interior.

Engineer Hill believed that the size of the farm unit should not exceed 20 acres, and could possibly be smaller, but cautioned that settlers probably could not be induced to take up land if the farm units were made too small. Forbes, whose experiment station had been in operation in the Yuma Valley since the spring of 1905, likewise advocated small farm units.[2] Forbes indicated that the long growing season and the rich alluvial soil favored intensive agriculture, which would prove immensely more productive per acre than larger farms growing crops similar to those found in the humid East. He also pointed out that there would be a rather heavy initial cost of leveling each farm parcel, perhaps as much as twenty-five dollars per acre, which in practical terms would restrict the average farmer to putting only limited areas under cultivation initially. And from a rural development standpoint, Forbes was convinced that intensive and diversified farming supporting many people would result in a wealthier and a more independent community, an important consideration at Yuma because of its isolated location distant from markets, and because of high transportation costs associated with shipments to those markets. Forbes's recommendation was to limit farm size to 15 acres, an even smaller unit than that suggested by Hill.

A. P. Davis was quick to endorse Forbes's assessment, for Davis had been advocating a farm unit ranging between 10 and 20 acres. In a communiqué to Forbes he expressed his appreciation for Forbes's corroboration of his own notions regarding farm unit size, noting that he was quite certain that "a small farm unit approximately what you suggest will be eventually adopted."[3]

The Yuma Valley settlers, particularly those who had filed land entries after the Reclamation Act became law, were intensely interested in the farm unit debate because they feared that the ultimate decision could affect the size of their landholdings. In June 1908, Oscar Bondesson, president of the Yuma County Water Users Association, received a letter from the secretary of the interior indicating that Forbes's recommendations would most likely be adopted, making the average farm unit 15 acres in size. After learning of this, a public meeting was held at Yuma to protest the impending decision, and a letter was drafted to the secretary of the interior expressing the opinion of Yuma Valley settlers regarding the size of the farm unit.[4]

The consensus among landowners in the valley was that farms entered prior to the implementation of the Reclamation Act should be restricted to no less than 160 acres, primarily because in many instances settlers had made substantial and permanent improvements on different subdivisions of their quarter-section tracts. For example, an average settler might build a house, put down a well, and build corrals and barns on the poorest land, and level and improve those tracts that were best adapted to farming. It would therefore be a hardship, it was argued, if one had to give up the improvements on one or more of these tracts. On the other hand, the settlers believed that the remaining farm units, those entered in the post–Reclamation Act period, should be at least 40 acres in size, primarily because most of those entries were of that size under the assumption that this would be the standard unit for reclamation homesteads. Part of the latter argument revolved around the condescending attitude among non-Indians toward the amount of land an average Indian family would receive (even before the 10-acre allotments were approved): "An average Indian family of five would have 25 acres. On this basis, 40 acres is surely not too much for a white man and his family," one settler declared.[5]

A more convincing argument against 15-acre farms was one that dealt with the type of farming that would be required of the settlers. A 15-acre unit would necessitate truck farming, since no other type of farming would support a family on such small parcels. But distance to markets and excessive transportation rates would make truck farming a risky endeavor for the average farmer. Results from Forbes's experimental farm were played down because they were obtained under ideal conditions, including expert supervision and management, attributes that few farmers on the project possessed. Hence, from the farmers' standpoint, a recommendation concerning farm unit size based solely on two years' experiments on Forbes's "play farm" were considered to have little relationship to reality.[6]

Before submitting his final recommendation to the secretary of the interior in the matter of farm unit size, director Newell also contacted project engineer Francis Sellew, who replaced Homer Hamlin in that capacity in the fall of 1906. Among Reclamation officials, Sellew was closest to the scene and was therefore most knowledgeable regarding local conditions on the project. In his response to Newell, Sellew was quick to criticize Forbes's proposed 15-acre farm units.[7] Whether 15 acres would be enough land to support a family in the vicinity of Yuma, wrote Sellew, depended on the meaning of the word *support.* If the level of support anticipated by the Reclamation Act involved not only the ability to make an adequate living but also the capacity to make provision for hard times, then the proposed 15-acre farm unit in his judgment was too small. And one factor that had not so far entered into the equation involved the quality of the land found throughout the Bard unit. The working assumption was that given enough water, all the floodplain land after it was cleared and leveled would be more or less uniform in its ability to support crops. Consequently, on many projects, including Yuma, a soil survey by Reclamation engineers was not completed until years after the project had opened. The one soil property that concerned Sellew most was the distribution of alkali deposits, which would make some of the 15-acre parcels, at least initially, worthless for cultivation. If the farm were large enough to contain a reasonable amount of fertile land that could be cultivated while the alkali portion was undergoing treatment, settlement of the project, according to Sellew, would be greatly enhanced.

Sellew also sided with the Yuma Valley settlers regarding Professor Forbes's experiment station; he felt it was really an ideal farm, managed by experts who thoroughly understood irrigation methods and who were not competing with other farms in marketing the farm's produce. Sellew suggested that the same kinds of returns realistically could not be expected from settlers who were less proficient in irrigation agriculture, and he also wondered about the effect of many competitors entering the market at the same time. Additionally, he pointed out that the results of the experiment station, although destined to be of immense value to the project, had been obtained over so short a period that the "uncertain factors of season, crop, and market have not been developed to an extent which justifies basing the farm unit upon them."[8] Sellew believed that a farm unit of 15 acres meant that this would necessarily be a project of truck farmers, a type of farming that required a sizable nearby market, a circumstance that did not exist at Yuma.

Sellew was convinced that most settlers would initially raise staple crops, such as cotton, alfalfa, and grain, and perhaps some livestock, and that 15 acres devoted to this form of agriculture simply would not support a family. He therefore recommended that "the minimum farm unit for this project should not be less than 40 acres of good tillable soil." It is unfortunate, however, that Sellew, in an attempt to bolster his argument, strayed from a rational view of the situation to invoke the standard "Indian" analogy. He insinuated that if the intention of the Indian Department was to give each Indian 5 acres of cultivable land, an Indian family of five persons would then get 67 percent more than that proposed for the support of a non-Indian family. "If 25 acres is needed to meet the primitive requirements of an Indian, surely 40 acres is not excessive for a white family," Sellew contended.[9]

Sellew discussed his opinion with Professor Forbes, and also gave him a copy of his letter, before forwarding it on to Newell. The following day, Forbes, in a letter to Newell, retreated somewhat from his original position advocating 15-acre farm units. Forbes conceded that the issue of financial returns under competitive circumstances was a viable one; it was entirely possible that the market for gardening enterprises was limited, as was the number of individuals who could quickly adapt to intensive irrigated farming. Nevertheless, "I still believe that the maximum development of this region must be highly intensive in character," Forbes wrote, "for the simple reason among others that it is physically impossible for a farmer with a family and a reasonable amount of hired help to care for more than a limited area."[10] Forbes presumed that this applied even to alfalfa, corn, and other less intensive crops that, if not closely cared for, would be taken over by Bermuda grass in a short time, leaving the land more difficult to reclaim than in its wild state.

On the other hand, Forbes realized that the people who would eventually populate the project would come from regions where much larger farms were the rule, and that they most likely would not readily adapt mentally to the concept of small farm units. Forbes concluded that there was significant doubt about whether a 15-acre farm unit would induce prompt settlement of the project by a desirable class of settlers, and that it would probably be better to err on the side of liberality rather than parsimony in this particular decision. He therefore suggested that if the 40-acre parcels recommended by Sellew were found to be too large, then portions of those farm units could be relinquished; this would be an easier adjustment to make than if the original farm unit was too small, requiring an increase in acreage through purchase afterward. Without completely giving up hope for smaller units, however, Forbes emphasized that if it appeared that there would be a strong demand for relatively small units by industrious people who would personally farm their land, then farm units of perhaps 20 acres would lead to a more rapid and complete development of the project.

Immediately after receipt of Forbes's letter, Newell wrote Sellew indicating that 40-acre farm units should probably be adopted in the Yuma Valley where many of the post-1902 entries were filed under the assumption that 40 acres would be the size of the reclamation homestead, but in an area like Bard, where the land had not as yet been filed upon, he suggested it might still be wise to try 20-acre farm units, which could be increased in size if they were not taken promptly.[11] Newell also recommended that two townsites be laid out, one in the Bard unit on the branch railroad between Yuma and Laguna Dam, the other on the main Southern Pacific line west of Yuma in the Indian unit. Newell's primary reason for recommending the establishment of these townsites involved the assumption that settlers would be more likely to accept smaller holdings if they were located close to a town.

However, in another communication written the same day, Newell informed Sellew that apparently there had been a misunderstanding regarding the original proposal to make the farm units 15 acres in size. Newell clarified by stating that the farm unit could be as large as 40 acres, but that it should contain at least 15 acres of good irrigable land relatively free from alkali. In a roundabout way, Newell was gradually beginning to accept Sellew's position regarding farm unit size. Sellew, in his reply to Newell, was diplomatic: "While I feel pretty confident that 40 acres will be necessary for the support of a family within the meaning of the Act, I agree with you that the Indian Reservation is the proper place to experiment in this matter and believe it would be wise to try the 20-acre unit. The size could be readily increased, as you suggest, if the future showed such action to be necessary." Sellew also supported Newell's suggestion regarding the general location of the townsites: "These localities will be investigated and the towns properly laid out preparatory for settlement."[12]

Three months following this exchange between the project engineer and the director of the Reclamation Service, Newell and the secretary of the interior collectively made a decision regarding the size of farm units in the Bard district, and Sellew subsequently was instructed to prepare farm unit plats accordingly. Farm units were to be surveyed in 20- to 40-acre

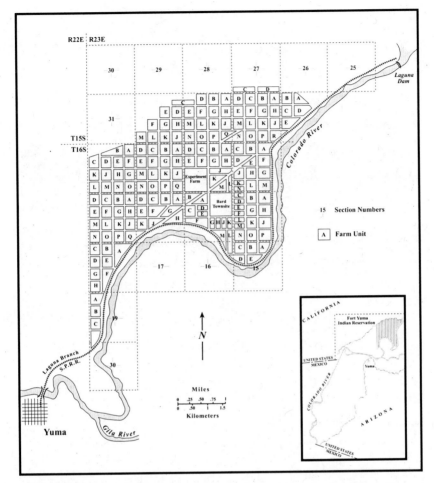

MAP 6.1. Farm unit survey in the Bard district. (Based on "Official Farm Unit Plat—Indian Reservation—Yuma Project," File 560-A1, Box 1101, National Archives, Washington, D.C., RG 115)

parcels, with the area of each unit broadly defined as being dependent upon the circumstances peculiar to it. Sellew had previously requested that a soil survey be completed prior to the survey of farm unit plats so that farm unit size could be based on the quality of the soil, but no action was taken by the service on this matter. Hence, Sellew interpreted his instructions liberally, for 153 of the 173 farm units surveyed were 40 acres in size.[13] The remaining 20 farm units were 20-acre parcels, most of which were located adjacent to the quarter section reserved for the townsite of

F I G . 6 . 1 . Office of the agricultural experiment station at Bard, 1926. (National Archives Photo, courtesy of the Yuma Area Office)

Bard (map 6.1). Immediately to the northwest, another quarter section was reserved for an agricultural experiment station, to be equipped by the U.S. Department of Agriculture in cooperation with the Reclamation Service, so that advice could readily be disseminated to farmers located within the Bard district (fig. 6.1).

It was not until the spring of 1910 that the Reclamation Service was ready to first turn water into the reservation canals to irrigate farm units located in the vicinity of Bard. A few months earlier, in January 1910, a public notice was issued announcing the intention to open for irrigation 6,500 acres of land in the Bard unit. When the public notice began to appear in local newspapers, the Imperial Investment Company attempted to divert interest toward the Imperial Valley by publicizing through the circulation of a flier the availability of land in the Imperial Valley. The handbill queried: "Why waste a year? Why not plant your crop at once? Why not locate where your market is at your door and calling for more than you can produce? Why make your women pioneer when you can locate in an established community? Why not get your land already graded, ditched and bordered, ready for planting? Why not locate at Brawley or Westmorland in the Imperial Valley? Why not investigate this and see how you can do it at a smaller cash outlay than by taking up land in the Yuma Project?"[14] In spite of the competition emanating from the Imperial Valley, considerable demand was anticipated for Yuma Project lands, and the task now was to determine how the farm units in the Bard district should be made available to settlers.

The Lottery

It was the General Land Office, not the Reclamation Service, that disposed of farm units within federal irrigation projects. At the end of February 1909, Newell informed the secretary of the interior that the lands in the Bard unit would most likely be viewed by the public as containing great potential value and urged him to conduct a drawing or lottery to determine who would be given priority to enter each farm unit. No immediate action was taken on this matter, but the General Land Office had always assumed that everyone should have an equal chance to obtain government land and that no screening of entrymen should be required to determine the likelihood of their succeeding as farmers. One year later, as the time for the opening of the Bard farm units approached, the *Los Angeles Times* published an article titled "Line Hitters Standing Pat: Yuma Land Seekers Still in Aggressive Mood; No Change in Department Plan for Filings; Indications Are for Fierce Rush at Opening." The *Times* reported that 150 potential entrymen had already begun to form a line in the hallway outside the Los Angeles district land office located on the fourth floor of the Chamber of Commerce building in downtown Los Angeles. "It doesn't take a vivid imagination," noted the *Times*, "to picture the scene that will be enacted in the narrow quarters of Receiver Robinson when several hundred eager, impatient men are pushing into the doorway."[15]

In order to maintain control of the growing crowd, the police were handing out numbers to those standing in line, in spite of the announcement that the register and receiver of the land office had planned to secure a large hall, possibly the Shrine Auditorium, where the names and addresses of all applicants would be recorded. Once the requisite information was carefully logged, the doors of the hall would then be flung open and the rush would begin. Since land office policy did not prevent simultaneous entries for each of the 173 farm units, it would be the responsibility of the register and receiver to sort out the complications resulting from the chaotic scene. One month earlier, in fact, concerned about the potential confusion and the conflicts that would develop from such tumult, the land officers in charge of the Los Angeles district land office jointly had sent a telegram to the commissioner of the General

Land Office proposing that the opening be done by drawing. The lottery would have popular appeal, they reasoned, because it would give all potential entrymen a chance to share in the benefits of the program. Given the probability of a frenzied rush for Yuma Project lands as reported in the *Times,* the secretary of the interior gave his approval and instructed the General Land Office to conduct a drawing for the 173 Bard farm units.[16]

At nine o'clock on March 1, some 2,800 people gathered at the Shrine Auditorium in Los Angeles to receive I.D cards that entitled them to make a formal entry for land located within the Bard unit.[17] Land selections by potential entrymen had to be filed at the district land office in Los Angeles prior to March 20, and all entries made during that time frame were considered to have been filed simultaneously. Demand was indeed intense, as more than 1,726 entries were filed on the 173 parcels.[18] The number of applicants per parcel ranged from a low of 2 on 6 different farm units to a high of 151 on 1 unit. Farm units located around the margins of the Bard district tended to receive the fewest number of applicants, while the frequency of applicants increased with proximity to the townsite of Bard and the experimental farm (map 6.2). In fact, the most popular farm unit by far was a triangular-shaped parcel wedged between these two reserved plots. The panhandle portion of the Bard unit located in the extreme southwest corner was also attractive to prospective entrymen, primarily because it was situated only a couple of miles from Yuma. In contrast to the general belief of the Reclamation Service, the fact that this section lay immediately adjacent to the Quechan allotments in the Indian unit apparently had no bearing on the desirability of these units.

On March 22, the drawings began at the land office. Twenty-seven parcels were disposed of on the first day of the drawings, and the drawings continued through March 28, when the last of the 173 farm units was transferred to private ownership. The procedure was to admit those individuals who had filed on a given farm unit. Each wrote his name on a card, sealed it in an envelope, and dropped it into a box. One of the prospective entrymen was blindfolded and asked to pick an envelope from the box. The successful entryman, the one whose name was drawn, was given ten days to make his first year's construction charge payment, amounting to $220. (This was the charge on a 40-acre farm at a rate of $55 per acre spread over ten years.) There would also be imposed an operation and maintenance charge of $10 per acre, and, in accordance with the statute permitting the allotment of the reservation, a charge of $10 per acre

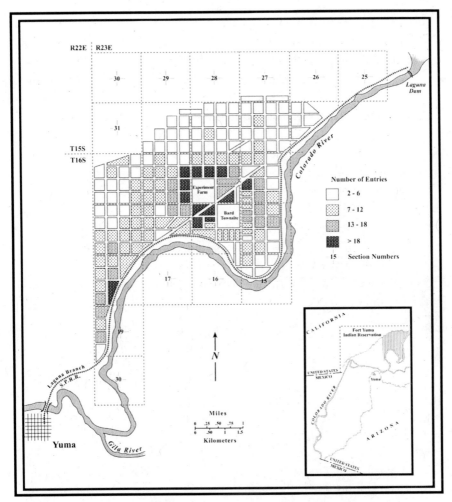

MAP 6.2. Number of land entries by farm unit in the Bard district. (Compiled from "Farm Only for One in Ten," *Los Angeles Times,* March 23, 1910, File 153-11, Box 1087, National Archives, Washington, D.C., RG 115)

would be assessed on the Bard lands to assist in the reclamation of the Quechan allotments, or to be used from time to time to benefit the tribe in other ways. The latter two charges would amount to an additional $2 per acre per year, or $80 per 40-acre parcel. A total of $300, therefore, was required as a down payment on the 40-acre plots; the 20-acre farm units required half as much.

In some instances, where only a small number of applicants competed for a particular farm unit, the parties met outside the land office and arranged a settlement so that one of the applicants received the farm without contest. In other cases, after the drawing on a parcel was completed, one of the losing applicants would offer $500 to $600 for the coveted farm unit. In most cases, however, the lucky applicants were reluctant to sell their claims.[19] No screening of applicants was done to determine their financial ability to take up irrigated farming, or even to ascertain their farming background, nor were the applicants required to make a personal examination of their desired farm units prior to filing their entries. Some years later it was said that the result of opening the land by lottery to all comers was that "carpenters, painters, pool sharks, doctors, clerks, [and] speculators" were the fortunate persons, and that "no farmer was lucky enough to get a unit."[20] The sad fact is that without at least a preliminary screening of the applicants, combined with the lack of a soil survey of the 173 farm units, the settlement policy devised by the Interior Department and the Reclamation Service was seriously flawed from the beginning.

The Bard Townsite

In April 1906, Congress gave its approval to the Town-Site Act, which provided the Reclamation Service an opportunity to build model communities, something that it seemed eager to do.[21] Soon thereafter, Newell adopted a general town plan for Reclamation Service projects, one that was first applied to the townsite to be located in the Bard district. The plan called for eight broad avenues converging at the center of town; four were oriented in the cardinal directions, while four ran diagonally. Perhaps to emphasize the town's new priorities, a school rather than a business district was to dominate the center of the community (map 6.3). Because of the area's extreme summer temperatures, Newell suggested that Bard's central plaza be well shaded, with trees also lining the avenues radiating out from the center; in order to benefit from well-shaded residential areas, though, residential streets should not be too wide. Approximately 50 percent of the receipts from the sale of town lots would be used for municipal improvements and to supply water to the town. Instead of offering all the lots in the townsite for sale at one time, both business and residential lots should be placed on the market in accordance with demand. This would prevent speculation in town lots and allow the Reclamation Service to control the development of the town.[22]

TOWNSITE OF BARD

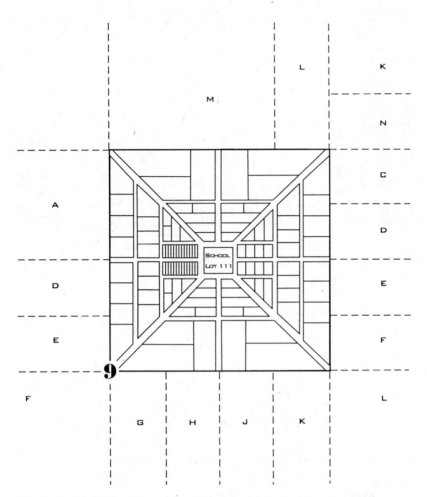

M A P 6.3. F. H. Newell's proposed design for the Bard townsite located in section 9 of the Bard unit. Farm units (*designated by letters*) surround the town. Scale is 1 inch = 800 feet. (Based on "Townsite of Bard," File 560-C, Box 1105, National Archives, Washington, D.C., RG 115)

F I G . 6 . 2 . The Bard general store still stands on the former Bard townsite. (Photo by the author)

Like Reclamation settlement policy generally, the townsite scheme for Bard was flawed. In preparation for settlement, the 160-acre townsite was surveyed and subdivided in 1911, but five years later no lots had as yet been offered for sale because there simply was no demand on the part of the settlers to establish a town at this point. It will be recalled that Newell's primary rationale for establishing a townsite at Bard was to influence the size of farm units, keeping them relatively small. Yet all but twenty of the parcels were surveyed to conform to the standard-size reclamation homestead of 40 acres. The founding of the Bard townsite, therefore, had nothing to do with economic necessity, and the townsite experiment resulted in failure. By 1914, because of the experience at Bard, the Reclamation Service had lost all interest in community planning.[23]

In reality, the Bard farm units lay in the shadow of Yuma, an established farm service center conveniently located on the transcontinental Southern Pacific rail line. Because of the lack of interest in the sale of Bard town lots, the Reclamation Service tried to lease them at a nominal fee. As of 1916, only one lot had been leased—to E. F. Sanguinetti, the most prominent merchant in Yuma—on which a general store was established (fig. 6.2). The remainder of the townsite stood unimproved, having grown up to a thick jungle of arrowweed, mesquite, and cottonwood. In 1917 the decision was made to scale back the Bard townsite to 40 acres, and to dispose of the excess 120 acres in three 40-acre farm units.[24] Because of the Bard experience, the townsite within the Indian unit, which was to be known as Powell, was never even surveyed. Unfortunately, the fate of these townsites was a harbinger to the host of difficulties that would eventually plague the reservation division of the Yuma Project.

7

Yuma Valley Travails

The Yuma Valley presented a set of problems for the Reclamation Service different from those found in the Indian and Bard units across the river. The majority of land in the valley had already been filed upon and was in the process of being transferred to private ownership when the Yuma Project was approved. It was widely assumed when the federal reclamation law was implemented that the Reclamation Service would focus its efforts on bringing unsettled public lands under irrigation, but there were many instances when the greatest opportunities for the development and settlement of a federal irrigation project were afforded by areas composed of a combination of public and private lands. In fact, there was a considerable amount of private land included in nearly all early irrigation projects. The Yuma Project was no exception. Both the Imperial and the Yuma valleys represented areas largely in private ownership within the general purview of the Yuma Project. Although federal attempts to embrace the Imperial Valley were unsuccessful, the Yuma Valley was more critical to the immediate success of the project. The area encompassed by the Indian and Bard units included approximately fifteen thousand acres of irrigable land, whereas the Yuma Valley included more than three times that amount. Without the endorsement of Yuma Valley settlers, the development of the Yuma Project would not have been feasible.

After the Yuma Project's approval, a sense of optimism swept over the Yuma Valley. Valley residents believed that federal intervention in the development of irrigation would bring an end to the difficulties that they had theretofore struggled with in their attempts to bring the valley under irrigation. But their enthusiasm soon turned to disappointment as an extraordinary series of problems befell the region in the early years of federal involvement. Initially, the settlers of the Yuma Valley began to clear and level their land and plant crops, anticipating that good prices would be paid for farm produce during the construction phase of the project.[1] But the floods on the Colorado River below the Gila confluence in the spring and early summer of 1905, which inundated a large portion of the Yuma Valley, squelched this activity by swamping settlers' houses, drowning out their crops and covering fields with a thick deposit of silt, and rendering the two gravity canal systems in the valley useless. Only the Farmers Pump Canal, which covered the higher east-side lands in the Yuma Valley, was able to continue to deliver water to those parcels that were not flooded. In a very short period of time the fruits of the labor of Yuma Valley farmers were destroyed, impoverishing many of them (fig. 7.1).

Floodwaters on the lower Colorado River also resulted in significant delays in the completion of Laguna Dam, the cornerstone upon which the entire project rested. With the Colorado River out of control and the fear that its riverbed would be eroded to a depth that would make it impossible to divert it for irrigation, active prosecution of the work at Laguna Dam was suspended until early 1907, after the Colorado River was brought back into its channel. This delay, and others that followed, eroded the early confidence in the Reclamation Service held by Yuma Valley settlers. Having endured the privations of life on the frontier without much success for several years, they had elected to place their trust in the service to assist them with the completion

FIG. 7.1. An abandoned Yuma Valley homestead in December 1906 with highwater mark resulting from the 1905 floods. (National Archives photo, courtesy of the Yuma Area Office)

of the venture that they had begun. But to their dismay, Laguna Dam was finished two years behind schedule, and the Colorado River siphon, the link that connected the Yuma Valley to the rest of the project, was not ready for service until the summer of 1912. Although the difficulties experienced in the Yuma Valley during the eight-year span between approval of the project and completion of the siphon can partially be blamed on the weather-related events of 1905 and 1906, bureaucratic loitering is also responsible for causing much suffering in the Yuma Valley.

Canal Controversy

Approximately five thousand acres out of a possible fifty-five thousand had been reclaimed in the Yuma Valley when the Yuma Project was approved in 1904.[2] The major canal operations in the valley supplying water to farmers included the Farmers Canal and the Farmers Pump Canal, which had joined to form the Yuma Valley Union Land and Water Company, a cooperative enterprise operated by the settlers, and the corporate venture of the Irrigation Land and Improvement Company, also known as the Ludy Canal Company. In September 1904, supervising engineer Lippincott informed Newell that the Farmers and the Ludy Canal systems covered essentially the same lands in the valley and that both were frequently troubled by floods and low water, often placing them out of commission, but that it was very important for the sake of the farming interest of the Yuma Valley that at least one of these systems be efficiently maintained until the completion of the headworks and the main canal of the Reclamation Service. Lippincott stressed that it would be of benefit to the Reclamation Service to help sustain this farming district during the construction phase because it could serve as a nearby base of supplies. Three months later, however, Lippincott informed Newell that "the farmers under the old canal systems have ceased adequate maintenance or extensive repairs of their canals." The presumption among the settlers was that now that the government had entered the field, after the overwhelming majority of them had signed water contracts to purchase water from the Reclamation Service, the canal systems of the Yuma Valley would thenceforth be maintained and operated by the service, which now would supply them with water. Given this unforeseen situation, Lippincott recommended "prompt action for the relief of the settlers in the Yuma Valley and for the construction of the Yuma Project."[3]

An important consideration, then, which eventually turned into a seething controversy after the government's intercession, was which of the canal systems should be purchased and maintained during the period of construction. There existed a conspicuous government bias against the Irrigation Land and Improvement Company because it was understood that this corporation, formed by capitalists from the state of Washington, had "invaded the territory" of the Union Land and Water Company.[4] Negotiations, therefore, began immediately to transfer the farmers' canal operations to the United States for use during the time that the Yuma Project was under construction, and as early as October 1904, the Yuma County Water Users Association submitted a proposal to the government for the purchase of the farmers' irrigation systems in the Yuma Valley.[5] But partly because of the damage attributed to the rampaging Colorado River during this period, a delay of nearly three years occurred before the Reclamation Service purchased the farmers' canals and began delivering water to the Yuma Valley.

Meanwhile, with the canal systems out of commission, substantial suffering among valley residents ensued, as did bitterness toward the Reclamation Service, while the government's apparent lack of interest in the irrigation works of the Irrigation Land and Improvement Company caused an outcry among those who held an interest in this venture. The tables had been turned. Now the Reclamation Service had invaded the territory of this enterprise, and the corporation cried "foul," claiming that it was being unfairly treated by Reclamation officials.

As early as March 1905, the attorneys for the Irrigation Land and Improvement Company, one of whom was George Turner, a former senator from Washington, wrote the secretary of the interior in an attempt to gain some recognition by the government of the corporation's efforts to develop the Yuma Valley, and to advocate a government takeover of the enterprise after making compensation for its actual value, which the corporation estimated to be five hundred thousand dollars. It was maintained that the only other competitor in the valley, the Yuma Valley Union Land and Water Company, had for the previous three years actually obtained water from the Ludy Canal system. This was stretching the truth, but the ultimate aim of the Irrigation Land and Improvement Company was to gain a monopoly on irrigation in the valley. Turner noted in his letter that the corporation's venture was made with the intention of eventually having all the land in the valley tributary to its enterprise; otherwise, the investment would not earn adequate returns. But with the government's intervention, the venture would thenceforth be confined to land owned by farmers who had not signed contracts with the Reclamation Service, and it was now impossible for the

Irrigation Land and Improvement Company to compete with the service for future customers. Reimbursements from the existing limited amount of irrigated acreage would not even cover operation and maintenance expenses, let alone generate a profit. "All we ask is to be fairly recouped," pleaded the attorneys, "and we will cheerfully step aside."[6]

The secretary of the interior forwarded this appeal to Newell, who in turn requested that Lippincott investigate the matter. Two months later, in May, a five-member board of Reclamation engineers, which included Lippincott, submitted its report to Newell, recommending the purchase of the farmers' canals, which should be repaired and operated under the terms of the Reclamation Act. No recommendation whatsoever was made regarding purchase of the Irrigation Land and Improvement Company's canal enterprise because it was determined that the Ludy system was both a physical and a financial failure and that the reimbursement figure that its attorneys had suggested was absolutely unreasonable. The board also endorsed the early completion of a twelve-mile levee to protect the Yuma Valley from flooding. The construction of the levee would have the ancillary benefit of providing employment to many who had lost their homes during the recent spring floods. "From the human point of view," declared the engineers, "the settlers of the Yuma Valley are deserving of the aid so long promised and, up to the present, withheld." The board concluded, "[The settlers] have by their labor and money practically built the canal systems now in operation, and received but little in return. Now, with their lands devastated by floods, their crops destroyed, without means, and disheartened by the delays in prosecuting work on the Yuma Project, they are certainly deserving of aid, through the construction of works which must soon be built in any case, and through the results of proper management of their canal systems."[7]

In late November 1905, the Irrigation Land and Improvement Company filed a bill of complaint against the secretary of the interior in the U.S. Supreme Court, alleging that the United States in the context of the Yuma Project was in competition with the company, that the secretary was exercising powers not conferred by the Reclamation Act, and that the act was unconstitutional, and appealing for an injunction restraining the secretary of the interior from further prosecuting the work on the Yuma Project.[8] In May 1906, a demurrer was filed by the government, and in the following month the court ruled that the subject matter of the complaint was not within the jurisdiction of the court, ordering that the complaint be dismissed for want of jurisdiction, at plaintiff's cost. An offer was then extended by the Irrigation Land and Improvement Company to submit to arbitration regarding the value of its property, or to a decision of the Court of Claims, both of which were denied.[9]

In reality, the government had little interest in negotiating with the Irrigation Land and Improvement Company because the latter had entered the field later than, and had raided the territory of, the farmers' canal systems, and also because it had encouraged its water customers to refuse to sign contracts with the Yuma County Water Users Association until it could dispose of its irrigation works.[10] And the appropriation of a "navigable" stream for irrigation purposes by this corporation, technically, had been illegal anyway. Unlike the California Development Company, which was able to evade this constraint by seeking concessions from Mexico, the location and general slope of the Yuma Valley toward the south and west precluded a similar move by the Irrigation Land and Improvement Company.

The Petition

As conditions in the valley continued to worsen because of the reluctance of the Reclamation Service to take control of any of the canals, a new group of water users, calling itself the Consolidated Water Users Association of Yuma Valley, was organized to urge the government to take over and operate one or more of the valley's canal systems by lease, purchase, or otherwise so that farms in the valley could be supplied with water. Some settlers defected from the Yuma Valley Water Users Association to join the Consolidated Water Users Association in order to openly criticize the government's intransigence. Since the Yuma Valley Water Users Association was organized under the auspices of the Reclamation Service, its board remained silent concerning government policy; the Consolidated Water Users Association, on the other hand, provided a forum whereby concerned citizens could air their grievances.

One of the first actions taken by the newly formed organization was to circulate a petition among its members memorializing the Reclamation Service to do something to relieve the now pervasive suffering in the valley.[11] The petition carried the signatures of a surprising number of residents, 307 in all, and was forwarded by George Turner, former senator and attorney for the Irrigation Land and Improvement Company, to Interior Secretary E. A. Hitchcock, in March 1906. The petition summarized the views of residents regarding the economic plight of the Yuma Valley, brought on by the inaction of the Reclamation Service. The petitioners claimed that the Yuma

Valley was "thrifty and prosperous" until the time that the Reclamation Service entered the field and presented its plans for a unified irrigation system, and that they gave their hearty and undivided support, believing in the service's promise that the irrigation systems of the valley would be purchased by the government in order to furnish water to the settlers pending the completion of the project.

The petitioners placed the responsibility for the current situation squarely on the government's shoulders because the water-right contracts that the landholders were required by the Reclamation Service to sign eliminated the possibility of private competition, thereby destroying private irrigation initiatives. In other words, the owners of the irrigation systems saw no reason to continue maintaining their canals since their customers had been taken from them, and the existing canal companies turned their attention and energies to the adjustment of their claims with the Reclamation Service, and not to furnishing water to the lands linked to them. This situation resulted in a virtual standstill in the development of the valley, and in some cases a return to desert conditions as destitute settlers began to abandon their farms.

From the settlers' standpoint, their troubles all began with the substitution of the government's water system for those that already existed in the valley, with no one to maintain and manage the old systems during the time that the new one was being constructed. Not only had this unfortunate condition "wrecked the canal companies of the Valley," wrote the petitioners, but unless relief was provided by the Reclamation Service by operating one or more of the canals, "our homes will be reduced to a desert condition before the completion of the Government system."[12] The more preferable option, argued the farmers, would be to provide the necessary relief, which in turn would result in the expansion of cultivated land in the valley, thereby preparing settlers to use the water from the Yuma Project when completed. The petitioners professed that had they known the Reclamation Service would pursue the present course of inaction, they would not have signed their lands under government contracts.

The following month Lippincott forwarded to Newell a copy of the petition that he acknowledged had been "extensively signed" in the Yuma Valley. Even though the petition was supported by a large number of water users in the valley, many of whom were among the champions of the Yuma Reclamation Project, Lippincott suspected that the petition was placed in circulation by J. E. Ludy, vice president and general manager of the Irrigation Land and Improvement Company. Lippincott admitted that this was an ingenious move on the part of the canal interests in the valley, who were apparently "trying to sell us their worthless property."[13] He based this assertion on the fact that the signed petition was sent to former senator Turner, who was representing the case of the Irrigation Land and Improvement Company to the government. Members of the Consolidated Water Users Association, however, maintained that the petition was sent to Turner only because of their confidence in Turner's ability to properly and fairly present their case to the secretary of the interior.[14] Lippincott conceded that the farmers of the valley needed water badly. But he underscored the fact that all of the canal enterprises that had previously attempted to divert water by gravity from the Colorado River had been failures.

Lippincott informed Newell that it was the intention of the Reclamation Service to purchase one or more of the canals in the Yuma Valley at reasonable cost, and improve them, but the 1905 floods intervened, practically putting all of the gravity canals out of commission and making them of little value to the service. The surface level of the Colorado River had also fallen because of the problems downriver at the Imperial Canal heading, which compounded the situation by leaving the headgates of the gravity canal systems in the valley high and dry. Lippincott did acknowledge that negotiations were under way for the purchase of the one semiuseful canal enterprise, the Farmers Pump Canal, but no agreement had as yet been signed. He also contested the petitioners' supposition that the Yuma Valley was thrifty and prosperous before the service entered; rather, in his judgment, the settlers were "poor and in a struggling condition."[15]

In spite of the difficulties, Lippincott did not believe that property in the valley had depreciated in value. He theorized the contrary: that the mere presence of the Reclamation Service and the influx of contractors and laborers had resulted in an increase in property values, although he presented no firm evidence to substantiate that assumption. On the other hand, the Consolidated Water Users Association claimed that speculators were purchasing the homes of discouraged and impoverished citizens at half the price that they were worth at the time the Reclamation Service entered the picture.[16]

The Government's Response

Meanwhile, the secretary of the interior referred the Consolidated Water Users Association petition to Charles Walcott, director of the U.S. Geological Survey, who oversaw the Reclamation Service. Walcott, in turn, asked Newell to draft a response to the petitioners that would then be forwarded to the secretary of the interior under Walcott's signature.[17] Newell wrote that from the beginning the Reclamation Service had endeavored to make arrangements with the existing canal companies in order to include those parts of the systems that could be incorporated into the government's plan, and to encourage the owners to keep them in operation during the construction phase of the Yuma Project, primarily because it was apparent that once construction began the demand for agricultural products would exceed the ability of farmers to supply those products under existing conditions. Hence, it would not be reasonable for the Reclamation Service to deliberately put these systems out of business, for it would diminish the available source of supplies during construction, leading to increased costs, and because any further deterioration in the irrigation systems that could ultimately be utilized by the government would also result in increased expenses.

But Newell also pointed out that the Irrigation Land and Improvement Company, which had the largest system in the valley, claimed to have expended about a half million dollars in its development, while the evidence on the ground suggested that it should have cost but a fraction of that amount. Furthermore, the canal system had deteriorated badly since its construction. Newell informed his superiors that in April 1905, the board of five Reclamation engineers examined the conditions of the Irrigation Land and Improvement Company's canal system, and the following month reported that it had been both a financial and a physical failure, and that it was the board's opinion that it could not be materially improved in either respect. The board determined that the value of the system to the government was not more than twenty-five thousand dollars. Yet the company insisted that its property was worth five hundred thousand and that this amount was the only just compensation.

To bolster the Reclamation Service's position, Newell reported that the measurements made by the service prior to the 1905 floods indicated that the water supply furnished the farmers under the Ludy Canal system simply was not adequate for the proper irrigation of those lands. For weeks at a time during 1903 and 1904 there was either no water flow or such a small amount as to be wholly inadequate for the needs of the irrigators. Then, in 1905, after the Colorado River jumped out of its channel and began to flow westward into the Salton Sink and the river began cutting back its grade, the river channel became so low at the intake of the Ludy Canal that it was impossible to divert water into the company's system except when the river was at flood stage. And since the company had provided no adequate means of protecting its irrigation system against the effects of the annual floods by building levees, the unusually high water of the previous two years had practically destroyed the usefulness of its works. The other canal systems in the Yuma Valley suffered a similar fate, and they too had not been able to furnish a water supply sufficient to raise crops as in former years. Newell therefore concluded that the impairments to the irrigation systems were the result of natural conditions and not due to the actions of the Reclamation Service.

Newell also hinted that it was his belief that the majority of petitioners were endorsing the position that the government take over the canal systems in order to furnish adequate supplies of water for irrigation, rather than supporting the allegation that the Reclamation Service was responsible for the conditions that now prevailed in the valley. After all, if irrigators had been satisfied with the supplies they were receiving previously, why would they have overwhelmingly supported government intervention? The experience of the Reclamation Service in other projects, Newell recounted, indicated that wherever there was an existing system that was adequate, irrigators would refuse to sign obligations to apply for water rights under the Reclamation Act. Newell argued that the operations of the Reclamation Service could not have been destructive of successful enterprise, because it was only in those cases where the water service was inadequate that the water users under existing canals were willing to agree to take water from a government project. Newell, of course, failed to mention the fact that the settlers were indirectly coerced into accepting the government's plan for irrigation because their appropriation of Colorado River water technically was illegal. Finally, Newell concluded that the only reason one or more of the existing canal systems had not as yet been purchased by the government was because of the impossibility of obtaining reasonable terms. The cost of these systems had now become an even more important consideration from the standpoint of the Reclamation Service in light of the substantial cost overruns already incurred in the construction of Laguna Dam.

Walcott essentially rubber-stamped Newell's response by signing his name to a copy of the same letter, and forwarded it to the secretary of the interior. On May 29, 1906, Secretary Hitchcock replied formally to the Consolidated Water Users Association through former senator Turner. The secretary quoted passages from the above letter, concluding that the wreckage of the canal systems in the valley was due entirely to conditions over which the government had no control, and, in any event, it was not in the government's best interest to do anything that might injure these systems.

"The truth of the matter," Hitchcock quoted Newell via Walcott, "is that the gravity systems were physical and financial failures prior to the beginning of the government operations. The physical conditions have been so aggravated by natural causes that it is now apparent that there was no possibility of furnishing an adequate water supply by the repair of the gravity systems, and that the expenditure by the government for that purpose would be wasted."[18]

The secretary emphasized that the quickest and only efficient means of supplying water to the Yuma Valley was by the completion of the Yuma Project, because he was not given the power to acquire, or to take possession and to operate for temporary relief of its patrons, a canal system that could not be incorporated into the irrigation works constructed under the Reclamation Act. He then proceeded to point out that the petitioners had neglected to identify any irrigation system that could be included in the Yuma Project at a cost that could reasonably be reimbursed to the government in construction charges, or any system that could be operated if it were acquired. "It therefore presents nothing within my proper powers under the irrigation act," Hitchcock concluded.[19]

Some settlers in the valley, particularly those tied to the Ludy Canal, nevertheless were convinced that it was the methods adopted by government officials that resulted in the "physical and financial failure" of the canal systems. Although Ludy continued to hope for a government buyout, members of the Consolidated Water Users Association now were not necessarily asking that the government purchase the existing canal systems but that it simply take over and operate one or more of the canals so that water might reach their parched and impoverished homes. In fact, the majority of water users in the valley opposed the purchase of the Ludy system because this sum would have to be reimbursed by the settlers in the form of increased construction charges even though no part of the corporation's canals could be incorporated into the government's plans for the Yuma Project.[20] On the other hand, it was estimated that only about ten thousand dollars would be required to repair the farmers' canals and place them in serviceable condition.[21]

Several months after being shunned by the secretary of the interior, the Consolidated Water Users Association saw a potential government ally in Louis C. Hill, who in July had succeeded Lippincott as supervising engineer in the region. As long as Lippincott was still in charge, whose reputation in the valley was by now significantly tarnished, it was presumed that the water users would never be able to get a fair hearing. Hill was provided a copy of the petition sent to the secretary of the interior, and the water users in their cover letter emphasized that the results of inaction by the Reclamation Service were evident by the "deserted homes and closed business houses, [and] the depreciation of farm and town property," and they contended that this unnatural condition was due to the "unfair, unbusinesslike and unjust methods pursued by the Honorable J. B. Lippincott."[22] Their primary dissatisfaction with Lippincott involved his unwillingness to come to terms with the irrigation companies. With the resignation of Lippincott and the appointment of Hill, it was hoped that there might yet be a solution to the grievances confronting the settlers under the Yuma Project.

Hill also was provided a copy of Secretary Hitchcock's reply, which, the Water Users Association alleged, was spawned by "the brilliant but deceptive intellect of ex-Supervising Engineer J. B. Lippincott, whose blighting and aggressive hand must have dotted every 'i' and crossed every 't.'" It was actually Newell who provided the raw material for Walcott's response to the secretary of the interior, the substance of which was adopted in Hitchcock's letter to the Consolidated Water Users Association, although Newell had received considerable input from Lippincott, particularly in regard to the "physical and financial failure" assessment. But the Consolidated Water Users Association accurately surmised that it was hard to believe that Hitchcock gave the petition his personal consideration "to the extent so important a matter deserved, from the manner of reply we received to it."[23] In any event, it was apparent to the water users that Hitchcock based his reply on the representations of his subordinates.

In fact, the entire letter to Hill summarizing the manner in which irrigation interests in the valley had been mishandled by the Reclamation Service was written in an indignant tone. One such passage, for example, stated, "When we signed our land under your contacts we were supposed to return to our homes and become the consumers of water, and not the producers and deliverers, and your department's violation of its promise has brought want, poverty and destruction of property in our midst."[24] The letter, however, concluded with an apology over having to be so critical of the Reclamation Service, and assured the newly appointed supervising engineer of the water users' cooperation so that harmony and peace could be accomplished out of past discord and unpleasantness. But the confrontational attitude and lack of subtlety, which permeated the water users' letter, were probably not the best way to introduce their grievances to a high-ranking engineer in the employ of the Reclamation Service. Hill's sensibilities were no doubt offended, and he completely disregarded the water users' plea for help.

The highest-ranking officials of the Reclamation Service, the U.S. Geological Survey, and the Department of Interior had stonewalled the Consolidated Water Users Association. Having no success whatsoever in gaining concessions from the government, but nevertheless believing strongly in their cause, out of sheer desperation the association, on January 26, 1907, wrote to President Roosevelt: "It is with a sense of keenest regret," the letter opened, "that we feel ourselves at this time compelled to call your attention to the bad faith and broken agreements, to the citizens, but more particularly to the settler and land-owner under the Yuma Project, by maladministration of the Reclamation Act and by that Department."[25] The water users informed the president of their appreciation of the benevolent cause that motivated the framers and supporters of the Reclamation Act, and appreciating the president's sincere desire that the outcome of the act be beneficial to the homes and home builders of the arid region made their action painful but also imperative.

The letter to Roosevelt was relatively brief, with copies of the relevant correspondence attached that "will place you in a position to fully understand our differences, and make recommendations commensurate with the importance it deserves."[26] The letter closed with a statement of the association's desire for an amicable adjustment and an early reply. It is doubtful that Roosevelt ever saw the letter, however, for the assistant secretary to the president referred the matter to Walcott, director of the Geological Survey, who with Newell's help had originally drafted the government's response to the water users. The water users' grievance had gone full circle. It was apparent that the Consolidated Water Users Association would have to try some other tactic if their complaint against the government was ever to be resolved.

A Final Plea

Out of frustration, in July 1907, the Consolidated Water Users Association decided to appeal to public sentiment by publishing a slick fifty-two-page pamphlet, *The Unfriendly Attitude of the United States Government Towards the Yuma Valley, Arizona,* the purpose of which was to disseminate the facts, at least as this group of water users viewed them, to the public concerning the mismanagement and misrepresentations of the Reclamation Service in its work at Yuma. In addition to including reprints of the correspondence that preceded this publication, it also summarized the events that led to the controversy, explained why the canals of the Yuma Valley had not been operated, compared the prosperity of the Imperial Valley to the degradation of the Yuma Valley, included photos of abandoned farmsteads and affidavits of settlers under the Ludy Canal system attesting to their satisfaction with water deliveries prior to the intervention of the Reclamation Service, as well as miscellaneous articles from a variety of newspapers and magazines that had reported on the conflict. The pamphlet concluded with a call for an irrigation convention on August 31 in Sacramento, two days before the opening of the National Irrigation Congress at the same location, in order to discuss the operations of the Reclamation Service and to draft a proposition to be introduced at the Irrigation Congress memorializing the president to mandate a thorough investigation of the operations of the Reclamation Service.

The Yuma Valley Consolidated Water Users Association not only charged in their circular that the Reclamation Service had made false representations to the settlers in the valley in order to secure their signatures to water-right contracts but now also accused the service with gross misrepresentations regarding the cost of the Yuma Project. Because of the obstacles encountered in the construction of Laguna Dam, which was now expected to cost double the original estimate, it was projected that construction charges would ultimately run somewhere between seventy and eighty dollars per acre for valley irrigators instead of the thirty-five to forty dollars per acre that the Reclamation Service had more or less guaranteed when the Yuma Project was initially proposed. The call was published in circular form and sent to approximately five hundred western newspapers and to leaders of the National Irrigation Congress, as well as to government officials. It was addressed "To the friends of Reclamation of arid lands throughout the arid regions of the United States," and was directed toward all settlers under other Reclamation projects that had been treated in a similar manner by the government, as well as toward those who sympathized with such settlers, and encouraged them to ensure that their interests were properly represented at the upcoming National Irrigation Congress.[27]

Whether the circulation of the pamphlet and the threatened action by the water users at the Irrigation Congress were responsible for forcing the government's hand with respect to taking over the farmers' canal systems in the valley is not known for certain. What is known is that a triumvirate of outspoken critics of the Reclamation Service from the Owens, Imperial, and

Yuma valleys joined forces, and a resolution critical of government action was introduced but voted down by the delegates participating in the Irrigation Congress at Sacramento.[28] Efforts to gain the sympathy of the delegates were no doubt undercut when, in the previous month, the steam-pumping plant, which had been partially successful in supplying water to the upper part of the Yuma Valley, was purchased by the Reclamation Service. Steps were then quickly taken to purchase the farmers' gravity canal system that covered most of the remaining formerly cultivated area of the valley. A new concrete heading for this canal was completed in the spring of 1908, and a large gasoline-powered scoop wheel was installed at the heading to provide for irrigation when the Colorado River was low.[29] These improvements now enabled the service to provide a continuous supply of water to between seven and eight thousand acres in the Yuma Valley.

In spite of the secretary of the interior's claim that he lacked the authority to acquire and operate a canal system that could not be incorporated into the plans of a reclamation project, the purchased canals in the end were unsuitable in terms of both capacity and location to become a part of the new system; therefore, no additional expenditures were made on them other than those necessary for their temporary maintenance and operation. Still, truly efficient service could not be expected under this arrangement because the irrigated area was made up of isolated tracts, some located nearly twenty miles from the headgates.

The government sold water for one dollar per acre-foot delivered at the farm, which was the same amount charged when the farmers themselves operated the system. Even though the returns from the sale of water did not meet expenses, the Reclamation Service was reluctant to increase the charge because of the depressed conditions that prevailed in the valley.[30] The service now readily admitted that the delays had embarrassed many of the settlers financially and that any additional burden would result in more departures from the valley.[31] Before the completion of the inverted siphon beneath the Colorado River in the summer of 1912, approximately 275 water users were being supplied by the jerry-rigged system of canals now operated by the government (fig. 7.2). Under existing conditions the amount of irrigated acreage had achieved its maximum, totaling approximately eight thousand acres, most of which was devoted to the production of alfalfa (fig. 7.3).[32]

But there were still many potential water users in the valley that remained unserved between the time that the farmers' system went back online in 1907–1908 and when the Reclamation Service siphon was completed

FIG. 7.2. A Yuma Valley homestead in April 1909 located on the Farmers Canal, operated at the time by the Reclamation Service. (National Archives photo, courtesy of the Yuma Area Office)

FIG. 7.3. Second cutting of alfalfa on a ranch in the Yuma Valley located two miles west of Yuma, May 1910. (National Archives photo, courtesy of the Yuma Area Office)

four years later. In October 1909, Interior Secretary Hitchcock's successor, R. A. Ballinger, came to Yuma to inspect the recently completed Laguna Dam. In advance of Ballinger's visit a petition was drawn up and signed by the president and six other members of the Yuma County Water Users Association. The petition conveyed the concerns of the settlers regarding a rumor that had spread through the valley that reclamation funds were not available to complete the siphon. They contended that Laguna Dam was of no practical benefit to the farmers and settlers who were now on the land, and who had been waiting for water for years, unless the main canal and siphon were also completed that would convey sufficient quantities of water from the California side of the river to their homes in Arizona. They therefore urged the interior secretary to make money available for the completion of the Yuma Project and that the energies of the Reclamation Service concentrate on providing for the needs of the present settlers, rather than furnishing a supply of water for those who would take up lands on other projects that had not yet been opened to settlement.[33]

This select group of farmers reminded the secretary that there were now more than twenty projects in progress, with the Yuma Project among the first to be undertaken. As one of the earliest projects, Yuma should have a prior claim to reclamation funds, they believed. Since some of the more recent projects had been started where there were as yet no settlers waiting for water, there would be no personal suffering if work on the newer projects was delayed and more attention given to the completion of the Yuma Project. "Much needed relief would be given to a struggling and impoverished community of several hundred settlers and their families who have been living from hand to mouth waiting for the water for the last six years, without which nothing can be done except what is possible by means of an expensive and limited pumping system which only provides water for 7,500 acres out of a possible 53,000 acres," wrote the settlers. "No crops grow here without irrigation," they continued, "and the cessation of work upon the Project will throw the locality back very materially; in fact, it will amount to a calamity which will ruin the prospects and homes of hundreds of people who have shown their confidence in the Reclamation Service to carry out its work by making their homes here, and who still believe in the efficiency of the officials in charge as well as in the ultimate success of the Project itself."[34]

The petitioners brought up a viable issue: why would the Reclamation Service delay the completion of existing projects, particularly those such as the Yuma Project, which embraced an already settled population, while continuing to add new projects to its agenda? Had fewer projects been undertaken, the settlers reasoned, they would have been completed more rapidly, shortening the amount of time for the delivery of water, thereby reducing the number of destitute settlers. Ignoring the political pressures for launching so many projects so quickly, there was an *official* explanation for this apparent paradox. The Reclamation Service's rationale for the piecemeal completion of the projects involved the amount of time required before an irrigation system could be utilized to its full capacity. The service maintained that each irrigation project was a growing organism that would develop in stages, and it was therefore unwise or impractical to build and complete a system with hundreds of miles of laterals and attempt to operate and maintain it at full capacity during the earlier years when farmers were slowly getting their lands into productive capacity and establishing markets.[35] It was largely on this basis that the service explained the addition of new projects before earlier ones were completed. Unfortunately for the Yuma Valley settlers, their irrigation unit was located farthest from the point of diversion, Laguna Dam, and hence mere distance from the source, as well as the necessity of having to complete a structure that would siphon irrigation water under the Colorado River, caused the Yuma Valley to be the last to be supplied with water. Nevertheless, by building so many projects simultaneously, "the Reclamation Service was repeating its mistakes before it had a chance to learn from them."[36]

There also was the matter of completing the siphon, which met with substantial delays. It is true that the Reclamation fund, because of the overzealous approval of projects by the secretary of the interior, had by this time become overburdened. To alleviate the financial problem, Congress, in 1910, advanced a twenty million–dollar bond loan to the Reclamation Service to speed up construction and get water to settlers. But the delay in completing the siphon under the Colorado River had less to do with funding and more to do with an attempt by Reclamation Service engineers to perform difficult subaqueous work without the use of pneumatic methods.[37] None of this work was contracted out by the Reclamation Service, and the reluctance to adopt new and untried methods tended to slow progress on the siphon. The Yuma Valley water users by now were impatient; after waiting seven years for the project's completion, and living from year to year on borrowed money, in April 1911 they decided to band together to construct a temporary flume across the Colorado River that would supplement the water supplies of the existing canals.[38] They argued that few settlers would be able to

FIG. 7.4. The Yuma Valley levee in the final stages of completion with construction of brush fencing and matting, December 1907. (National Archives photo, courtesy of the Yuma Area Office)

FIG. 7.5. View from the California side of the intake and cylinder gate structure of the Colorado River siphon in June 1912, just prior to its being placed into service. (National Archives photo, courtesy of the Yuma Area Office)

survive another year if relief was not immediately forthcoming. The local Reclamation engineers, however, vetoed this idea because, according to the settlers, the flume, if successful, would discredit the original plan to siphon the water under the river.

A few months later, the laterals, which were to receive water from the siphon, began to be constructed in the Yuma Valley. Meanwhile, the Reclamation Service levees had served to protect the valley from the floodwaters of the Colorado for several years (fig. 7.4). Finally, on June 29, 1912, the siphon was finished, and in recognition of the urgency to relieve the desperate conditions in the Yuma Valley, it was placed into service on that same day (fig. 7.5).[39] For the first time, the canals in the Yuma Valley were filled with ample and regular supplies of Colorado River water, unburdened by its heavy load of silt, delivered by gravity from Laguna Dam. After years of frustration, the link to potential prosperity in the valley had finally been forged. It was now the hope of all concerned that the long period of suffering was over, and that a new era of irrigation farming had dawned in the Yuma Valley. Unfortunately, this hopeful development would not play out overnight.

8

Distress and Discontent

The distress and discontent that prevailed in the Yuma Valley spread to other parts of the project as a multiplicity of new problems arose once water from the Colorado River began to be turned onto project lands. Many of the complications encountered on the Yuma Project were similar to those found on other Reclamation projects, particularly those dealing with the environmental repercussions resulting from the onset of industrial irrigation as well as with the socioeconomic problems produced by federal irrigation and settlement policy. But other difficulties were unique to Yuma. It is apparent that federal policy makers and Reclamation engineers were oblivious to the complexities associated with arid land reclamation. Unrealistic expectations among federal agencies, and the conflicts that developed between them, also resulted in sluggish progress.

It did not take long before government officials and the settlers on the Yuma Project realized that the mere turning of water onto dry land did not necessarily result in reclamation, that the process was far more complex than simply completing the engineering works that made irrigation possible. President Roosevelt, in his message to Congress in 1901 advocating a national irrigation policy, admitted that early reclamation efforts by the federal government would "of necessity be partly experimental in character."[1] As we have seen, the events leading to the approval and completion of the Yuma Project clearly illustrate the experimental nature of federal reclamation, but this facet of government intervention became even more apparent after Colorado River water began to flow toward newly planted fields.

The Bard Debacle

Settlement and development of the Bard unit began on a hopeful note. Within a couple of months after the drawing for the 173 farm units in March 1910, nearly 40 settlers were in the process of clearing and leveling their farm units, and, by the end of August, 71 entrymen had established residence on their holdings (fig. 8.1).[2] But only about 425 acres were under crop and receiving

F I G . 8 . 1 . The one-month-old Hadley homestead located in the Bard unit of the Yuma Reclamation Project, May 1910. (National Archives photo, courtesy of the Yuma Area Office)

water at the end of the first season. By the end of the following year, all but one of the 173 farm units was occupied, and the area under cultivation now exceeded 1,700 acres; however, this represented an average of only about 10 acres of cultivated area per farm.[3] Because of the relatively slow progress, settlers soon realized that they would have difficulty meeting the second installment of their construction charges, due on December 1, 1910.

One way to resolve this problem involved the relinquishment of entries. The original unit holders could legally assign the right to their homestead to another entryman for monetary consideration as long as the relinquishment paper was filed simultaneously with the new entry. By the end of 1910, 62 of the original 173 farm units, or nearly 36 percent, had been relinquished (map 8.1).[4] Most of those who relinquished their units in 1910, however, probably never really intended to farm, but instead were looking to earn a fast buck at the expense of the federal government. In succeeding years 30 more farm units would be relinquished. Of the 92 total farm units relinquished, 21 were relinquished twice, 4 were relinquished three times,

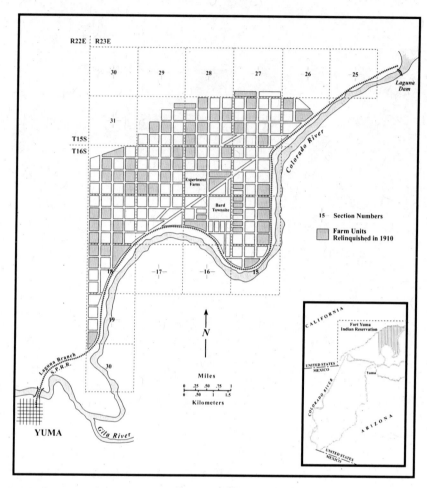

MAP 8.1. Farm units relinquished in the Bard district in 1910. (Compiled from Department of the Interior, Bureau of Land Management, Historical Indexes, Yuma)

while 1 parcel each was relinquished four and five times (map 8.2). Multiple relinquishments characterized the farm units located near the Bard townsite because they carried the highest speculative value, or those along the riverfront that turned out to be the hardest to work. Less than half of the 173 farm units at Bard were carried to patent by the original unit holders. The majority of the first settlers on federal irrigation projects relinquished their holdings because, according to Newell, they were inexperienced and expected irrigation farming to be much easier.[5]

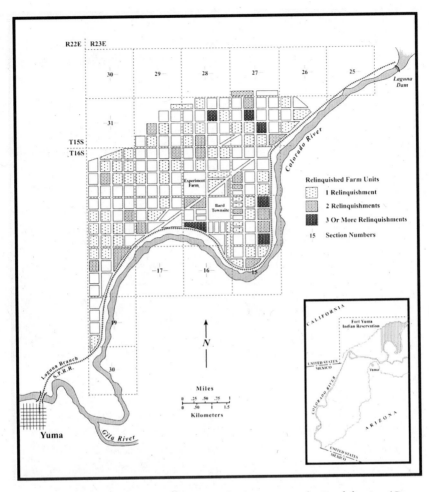

MAP 8.2. Frequency of relinquishments by farm unit in the Bard district. (Compiled from Department of the Interior, Bureau of Land Management, Historical Indexes, Yuma)

In reality, though, it was the cost of irrigation farming that forced many serious settlers off the land. Whether original or replacement entrymen, all homesteaders were suffering under the burden of having to pay the second installment of the construction charge. In November 1910, a petition representing a large percentage of the homesteaders in the Bard district was sent to the secretary of the interior requesting that the construction charges due in December be deferred for a year.[6] The settlers felt that since the first installment on construction was made at the time of entry, it was unrealistic to require payment on the second installment in the same year. They argued that the cost of clearing, leveling, and preparing the land for cultivation was far greater than anticipated, the average cost being approximately $65 per acre. And the inopportune time that the lands were opened prevented many of the entrymen from beginning residence until the extremely hot weather had passed.

Supervising engineer Louis Hill had anticipated these early difficulties even before the land was opened to settlement. In letters to Newell in May and June 1909, Hill had estimated that the cost of clearing the land in the Bard unit, building a house, purchasing farm implements, maintenance and operation charges, and basic living expenses would amount to $3,120 the first year. He correctly predicted that the addition of a construction charge of $55 per acre, plus another $10 per acre to reimburse the Quechan tribe for the land, if spread over ten equal annual installments, would require a considerable amount of money to enable even a good farmer to get part of his land in cultivation and to meet his payments the first year.[7] Consequently, the settlers, after several months' experience on the ground with little to show for their efforts, suggested in their petition that one way to ease their predicament would be to graduate the payments over the ten-year repayment period. Although the Reclamation Act stipulated that the government be reimbursed for all construction charges in ten annual payments, it did not specify that these payments had to be in equal amounts. If graduated payments were permitted, homesteaders could pay smaller installments in the early years, allowing them to use their capital for improving their holdings, and later in the ten-year cycle higher annual payments could be met from income from the land.

In June 1911, the plight of one settler, Richmond Wisehart, was expressed in a letter to Senator John D. Works, explaining that it was costing homesteaders two to four times what they were led to believe it would cost by government agents to clear, grade, and seed the land. He had only about 11 acres cleared and planted in alfalfa, and lamented that if required to pay the installment that was due the previous December, there would be no money left to continue expanding his cultivated acreage. Meanwhile, project engineer Sellew wrote supervising engineer Hill expressing his support for the case for relief being made by the settlers. He noted that most entrymen were unfamiliar with irrigation farming, and the cost of preparing the farms for cultivation had greatly exceeded their expectations. But more important, there simply was not sufficient time for them to put their farms under cultivation and obtain adequate revenue to meet the second payment due on December 1, 1910. Although Sellew did not advocate deferred payments, he did recommend that it would be "well to be lenient with them for the next year or two or until such time as they can reasonably get under cultivation an area that will make some return."[8]

The Department of the Interior, however, was in the process of devising an alternate plan for relief. In November 1911, a scheme was approved by the secretary of the interior for settlers on Reclamation projects, specifically the Yuma Project, to seek relief without implementing graduated payments by allowing for the voluntary reduction in the size of their holdings and applying the payment that had already been made at the time of entry to the smaller unit.[9] A settler, for example, could reduce his holding to 20 acres, or even to 10 acres, and relinquish the remainder to the government. Hence, the one payment made so far on a 40-acre unit could be applied to the portion that the settler retained. For example, on a 20-acre parcel, the payment would cover two years; on 10 acres, it would be good for four years. The primary motive behind this plan, of course, was to bring the average parcel size at Bard more in line with that originally hoped for by Newell and others in the Reclamation Service.

Director Newell, who was adamantly opposed to the graduation of payments on the Yuma Project, fully supported this alternate strategy for relief. He pointed out that the original intention, based on the best information available from agricultural and other experts, was to make these farms 20 acres. Newell was now more convinced than ever, considering the difficulties the settlers were experiencing on the Yuma Project, that 40 acres was too large an area to be successfully handled. "For this reason," he contended, "the relief . . . offered in the way of reduced holdings should be considered before the graduation of payments." But the settlers resisted this course of action, and, in March 1912, the secretary of the interior gave in to their demands, authorizing graduated payments spread over a ten-year period, with payments increasing from one dollar per acre to twelve dollars per acre between the second and the last installments.[10] In return, however, the construction charge was increased from fifty-five to sixty-six dollars per acre, presumably for the purpose of reimbursing the government for interest on the deferred payments. In spite of this increase, approximately one-half of the settlers, faced with the unacceptable alternative of having to reduce the size of their holdings or give up altogether, took advantage of the revised payment schedule.[11]

Notwithstanding the higher-than-expected costs associated with bringing the land under cultivation, the 1911 harvest season produced abundant
crops, especially of grain and alfalfa, even though only about one-quarter of
the total area was under production. Based on the first full season's returns,
therefore, prospects for the district appeared bright. But 1912 saw a significant turn of events. A long period of high water in the Colorado River lasted
during the months of June and July. The river lapped against the Reclamation Service levee that protected the Bard district from overflow, standing
eight to ten feet above the general level of the land for nearly six weeks.
Meanwhile, the sandy nature of the floodplain soils allowed the Colorado
River to pass freely beneath the protection levee. As a result, seepage un-
derneath the levee became severe. Approximately 4,600 acres out of a total
of 6,500 acres in the Bard district were affected by seepage. The farm units
located adjacent to the levee were hardest hit; for thirty to ninety days, 1,100
acres were totally submerged, while more than 860 acres became boggy.[12]
Water on the land did not recede until the river began to fall in August,
and then a heavy accumulation of alkali was left on the surface of more
than 2,500 acres. The result was a disastrous season of crop production in
1912. Compounding the problem during the year were periodic breaks in
the main irrigation canal, submerging many units because there was no adequate drainage channel to carry the water away.[13]

Even before the high flood stage of the Colorado was reached in the
summer of 1912, the water table had already risen to the point where the
construction of a drainage system was desperately needed. This was largely
because of the inexperience of the irrigators, resulting in their extravagant
use of water, combined with the naturally high groundwater plane in the
district. Work on the main drainage ditch in the Bard unit was therefore
begun in February 1912, but it did little to alleviate the seepage debacle
later that year. Ironically, the land that engineers and farmers were attempting to reclaim from aridity now had to be reclaimed from water.[14] While the
Reclamation Service remained silent about the seepage problem, in August
1912, O. B. Judd, a resident of the Bard district, wrote the secretary of the
interior informing him of the development of adverse conditions concerning
subirrigation and alkali on the farm units at Bard. He reported that two or
more feet of water covered many units and that all of the good land around
the townsite of Bard was ruined by "subing [sic] and salts."[15]

The secretary, in his letter of reply, tried to console Judd by assuring him
that the Reclamation Service was in the process of constructing drainage

ditches and other works to remedy the situation. Meanwhile, project engineer Sellew was asked by Newell to respond to the conditions described by Judd. Sellew admitted that seepage conditions were very serious at Bard, and that many settlers had suffered as a result. But he emphasized that the problem was partly compounded by the rise of the groundwater plane caused by the abnormal use of water by irrigators. Sellew believed that drainage would be effective only if accompanied by a reduction in water use. Sellew also revealed that he had received letters from forty settlers, as well as many other oral and written requests, seeking some form of relief because the events of the previous months made them unable to earn enough to meet their repayment obligations. In March 1913, temporary relief was approved by the secretary of the interior by reducing the payment that was due on December 1, 1911, to fifty cents per acre, with the balance due to be divided into two equal parts and added to the ninth and tenth installments.[16]

The problem of seepage during periods of high flood stage was noted back in 1903 when soil survey work was conducted on the California side of the river in the vicinity of Yuma, but its adverse effect on settlement was subsequently ignored as reclamation work progressed. When the Bureau of Soils published its results in 1905, about the time construction on the Yuma Project was beginning, it reported that from October through December 1903, the period covered by the survey, groundwater was within three or four feet of the surface on much of the overflowed lands. If these lands were to be protected by levees, the report continued, which would be required for successful agriculture, the level of the river would be artificially raised at flood stage, thus subjecting the groundwater to a greater head than before. Since the general formation of the soils involved a great bed of sand overlain and interstratified with layers of finer material deposited by the river, the water would rise rapidly in the subsoil, causing the soils nearest the river to become seeped and swampy. However, even on the silt loam, the most poorly drained soil type in the area, "with drainage to carry off the surplus of ground water during high stages of the river, almost any crop suitable to the climate could be grown," the report concluded.[17]

The advice of the soil scientists initially did not go unheeded, as the necessity for drainage was fully recognized in the original plans for the Yuma Project. In early April 1904, the board of engineers studying the feasibility of the Yuma Project wrote in their first report, "Much of the land is subject to overflow at high water, and it is consequently necessary to build levees to prevent this. Drainage channels are also necessary, and at times of high water these must be discharged by means of pumping. All these plans are contemplated in the estimates." And in a subsequent letter to the Yuma County Water Users Association providing a detailed description of the Yuma Project, chief engineer Newell stated, "Because these lands are so flat, and the level of the water in the ground so near the surface, it is considered necessary, for their permanent safe irrigation, to supply a drainage system. A main drainage canal has been designed to run through the central portion of the areas to be irrigated. . . . A pumping plant has . . . been designed to lift drainage waters over the levees during the flood period of the river to prevent the lands becoming water-logged."[18]

Another public statement demonstrating the importance of linking irrigation to drainage was included in a Reclamation Service pamphlet in October 1909, which provided a detailed description of the soon-to-be opened Yuma Project. The circular proclaimed, "No irrigation scheme can be successful without proper channels to control the ground water level and to remove the surplus water of irrigation."[19] And in 1910, unfavorable seepage conditions on tracts adjacent to the river had to be obvious to Reclamation officials when surveying parties laying out farm units were forced to wade through the riverfront tracts.[20] Hence, drainage was considered essential to the successful reclamation of this area long before the lands were thrown open to settlement. In fact, an appropriation of five hundred thousand dollars for drainage was included in the original estimate of the project cost.

But Newell in the years following the approval of the Yuma Project reversed his thinking regarding official policy toward drainage. His attitude now was that the need for drainage must become apparent before the large and expensive works would be constructed.[21] As a result, no drainage work was carried out on any portion of the Yuma Project until just prior to the 1912 seepage crisis. One month after a drainage system began to be constructed on the Bard unit in February 1912, Newell, in a letter to the secretary of the interior, clarified his position on drainage by insisting that "when the [Yuma] project was planned it was assumed that necessary drainage, except that required for the removal of surface waters, would be built by the farmers and not from the Reclamation fund. It has subsequently developed, however, that drainage on a somewhat extended scale will be necessary and that the irrigability of the lands can not be maintained unless these drains are provided by the Reclamation Service."[22]

Disregarding the findings of the government soil scientists years earlier, Newell presumptuously assumed that the need for drainage was largely due to the farmers' inept irrigation practices and disdainfully remarked that the Reclamation Service was having to take on this responsibility because of the farmers' unwillingness or inability "to perfect an organization to maintain the irrigability" of the project.[23] But Reclamation policy did nothing to encourage such an organization in previously unsettled areas such as Bard, since the operation and maintenance of the irrigation works remained the responsibility of the government rather than a water users' association. Now that the serious and costly need for drainage had materialized, Newell was apparently attempting to circumvent the original understanding involving drainage and lay the blame for this serious problem on the settlers.

The primary reason the provision of drainage by the Reclamation Service was downgraded from being essential to incidental to the project's completion no doubt was the excessive cost overruns already encumbered in building the Yuma Project. Newell did acknowledge that conditions on the project urgently required drainage and recommended that thenceforth the necessary cost of drainage be included in construction charges wherever such charges were yet to be determined. However, in the case of the Bard unit of the Yuma Project, where building charges had already been fixed, he suggested that drainage be included in the annual operation and maintenance charges assessed against the settlers.

Drainage on the Yuma Project, and especially the Bard unit, would be an expensive and difficult proposition. Under normal conditions, drainage of irrigated areas would be unnecessary during the early years of settlement because it would take several irrigation seasons before the water table would reach the level where surplus waters would have to be carried off. But conditions at Bard were different. Since a broad meander of the Colorado River washed against the levee protecting the Bard lands from overflow, this part of the project was most susceptible to a high water-table plane and underground seepage when the river was at flood stage.

There is no question, however, that the profligate use of water by inexperienced irrigators combined with the long irrigating season compounded the seepage problem. On the other hand, many farmers had no choice but to use extravagant amounts of water; some farm units, for example, were so sandy and devoid of organic matter that their water-holding capacity was extremely low, and it was only through frequent and copious irrigations that crops could be grown at all. This practice not only supplied sufficient water to thirsty crops but also increased the silt content of the soil, the only way to make the sandy lands productive. And the previously mentioned soil survey had reported that a high percentage of this area contained an "appreciable quantity of alkali."[24] Lands containing high levels of alkali could be made productive only through intensive washing and leaching out of the soluble salts by surface irrigation, but this would necessitate drainage adequate to permanently hold the water-table plane at a safe distance, at least five feet below the surface, to avoid the recurrence of alkali. Seepage, alkali, and sand would make drainage difficult because of the incredibly large quantities of water that would have to be handled by the system. There was also the matter of having to construct the drains in such a manner as to preclude the persistent and luxuriant growth of tules and other plants in open ditches that would otherwise destroy their usefulness.

The river was lower in 1913, and seepage of farm units was not so serious. During the year nearly 4,100 acres were under cultivation and receiving water.[25] But in April 1913, a 150-foot section of levee was washed out below Laguna Dam, causing some flooding. The branch line of the Southern Pacific Railroad, which had been built to haul materials for the construction of Laguna Dam, came to the rescue and helped repair the break quickly with the placement of rock protective works. Nevertheless, this event, combined with the fact that the land was left in such poor physical condition following the previous season's period of high water, led to the general failure of the 1913 crop season. And, of course, a recurrence of the 1912 episode could be expected each time the river reached high flood stage.

In order to avoid further disastrous consequences, project management was forced to focus on the immediate construction of a drainage system for the Bard lands. The original plan was to incorporate the natural sloughs and their tributaries coursing across farm units into a drainage system: "A comparatively small amount of inexpensive work in straightening and deepening these channels will fashion them into the most admirable system for the removal of surplus water," the Reclamation Service in 1909 naively conveyed to potential settlers.[26] But most of the natural drainage channels would ultimately be cut off by constructing canal and ditch embankments and road fills across them. Hence, in the summer and fall of 1913, a main open drain was dug through most of the Bard unit and 21,000 feet of newly laid underground tile drain was connected to it, and a pumping plant was installed just upriver from the siphon.[27] The tile drain, however was an immediate failure. Having been laid through cultivated fields, the pressure of heavy irrigated soil on the tiles caused them to collapse.

In 1914, the river was again higher than normal, causing 1,250 acres to become waterlogged, further reducing the productivity of the district. Subsequent to the failure of the tile drains, in the fall of 1914, an open drainage ditch about 200 feet distant from and parallel to the levee was partially excavated. But where its excavation was carried below the water line the material behaved like quicksand, and rapid sloughing off of the banks into the bottom resulted. After the expenditure of a considerable sum, the initial efforts to relieve the seepage conditions along the riverfront were aborted.

By February 1915, nearly five years after the opening of the Bard district, eighty-one farm units, or less than one-half the total, were growing crops on scattered and small areas, but not in sufficient quantities in most cases that would permit an acceptable standard of living. On an additional eighteen farms the soil was so sandy that the crops grown on them hardly paid the water charges. Only on about twenty-five farms, less than 15 percent of all farm units, were returns satisfactory and the unit holders prospering.[28] Virtually all of the remaining farm units, those located along the riverfront on seeped land, were in desperate condition.

Since most of the unit holders on the areas affected by seepage had spent their entire savings on clearing, leveling, and preparing the land for irrigation and on their survival during these difficult times, they were now nearly bankrupt. Although in August 1912 Congress approved the three-year Homestead Act, reducing the residency requirement from five to three years, allowing entrymen to prove up in a shorter period of time so that they could mortgage their farm units for much needed cash, this measure provided no relief to settlers who had taken up their farm units under the original five-year Homestead Act. But even if settlers were able to secure title to their land more quickly, banks were hesitant to loan money in the district because of its unfortunate reputation. In order to provide additional relief to settlers on the Yuma Project, as well as other projects where the plight of settlers was similar, on August 13, 1914, Congress gave its approval to the twenty-year extension act, allowing settlers who wanted to opt for this repayment plan to spread their payments on a graduated basis over twenty years.[29]

Under the new plan, farm unit owners would pay little other than operation and maintenance charges for the first five years while they were developing their land, and then payments would increase over the next fifteen years once the farm was producing. Almost immediately, 93 percent of all farm unit holders in Bard accepted the new repayment provisions, but their adoption of the twenty-year extension act came at a price.[30]

Acceptance meant the termination of the original contract of one dollar per acre for operation and maintenance charges; these charges would now be adjusted on an annual basis to meet actual costs, and they would have to be paid annually.

Unfortunately for the Yuma Project water users, operation and maintenance costs were excessive, much higher than those found on most other Reclamation projects, partly due to the expense involved in the removal of silt from canals. The desilting mechanisms on the project allowed for the removal of only about 25 percent of the sediment load carried by the Colorado River, mainly heavy sands but also a variable amount of silt in suspension (fig. 8.2). Beneficial amounts of alluvial material were allowed to infiltrate the system in order to provide fresh deposits of silt and nutrients to the irrigated soils. But the year-round growing season made it impossible to discontinue use of the canals for a sufficient length of time to allow them to dry out.[31] Meanwhile, a year's run of water resulted in a deposit of silt in the ditches from six to twelve inches in depth and an additional layer on the sides, necessitating annual cleaning.[32] The removal of the heavy mud, as a result, was expensive, averaging between $600 and $900 per mile.[33] And it was difficult work, mostly performed by Indians (fig. 8.3). Even though Laguna Dam and the settling basins were designed to remove some of the heavy alluvial load carried by the Colorado River, the reduced burden of silt that entered the delivery system was nevertheless choking it. Eventually, machines would do most of the canal-dredging work, resulting in substantial savings in time and money (fig. 8.4).

Test wells dug throughout the Bard district revealed that for approximately three miles back from the levee the underground water level rose and fell with the river. As a result, many of the farms at Bard that were originally productive were now deteriorating rapidly due to seepage and alkali accumulation. On about one-half of the Bard unit, and especially on the badly seeped area, the problem was no longer the ability to pay project costs but how to rescue a body of settlers from a hopeless situation. Many unit holders were obliged to make a living from occupations other than farming, some, ironically, becoming employed by the Reclamation Service. Others remained in farming by working for their more fortunate neighbors or by leasing Indian allotments, while a few unit holders who elected to do so were relocated to farm units in the Yuma Valley.[34] What was intended to become the showpiece of the Yuma Project was evolving into an economic and environmental nightmare.

FIG. 8.2. Silt trap at the head of the main canal at Laguna Dam. (Reproduced from F. H. Newell, "National Efforts at Homemaking," plate 9)

FIG. 8.3. Quechan workers operating "V" machine in ditch cleaning. (National Archives photo, courtesy of the Yuma Area Office)

FIG. 8.4. Ruth dredger cleaning a project lateral, June 1922. (National Archives photo, courtesy of the Yuma Area Office)

By 1915 the question was not whether drainage was necessary in the Bard unit but whether the agricultural value of the seeped lands warranted the high costs associated with drainage. One Reclamation engineer who examined the area at the time felt it ill-advised to attempt the complete drainage of the lands immediately adjacent to the levee, and that deep drains intended to lower and control groundwater should not be built any closer than six to eight hundred feet from the levee.[35] This would require the abandonment for farming purposes of a swath of land two to three hundred yards wide abutting the levee, although the eliminated land could still be used for pasturing livestock. The estimated total cost of the proposed drainage system was about $100,000. When added to what had already been spent in previous drainage attempts, the expenditures for drainage would total $250,000, or $40 per acre. Additionally, the maintenance of that part of the system composed of open canals would be costly because of the prolific growth of tules. The only effective means of handling this problem was to employ hand labor, again mostly Indian. Lining the open drains with concrete was a possible alternative, but this would entail even more expense, and the durability of such drains was questionable because concrete canal structures on portions of the Bard unit were already showing serious disintegration resulting from the action of alkali.

All parties agreed that an adequate system of drainage would be required to turn the Bard unit into a profitable farming district; the only question was who would pay for it. Understandably, the settlers felt they had been deceived by the government, not only over the issue of drainage but also because they had not been adequately informed about the character of their farm units and their suitability for irrigation. Since prospective settlers were permitted to file their entries without actually visiting and viewing the land, this further contributed to the perception of deceit. The Reclamation Service was noticeably embarrassed over the situation, not just at Yuma but where similar problems were found on other projects as well. Opposition to payments by water users on the projects became so serious that Interior Secretary Franklin K. Lane appointed local boards of review for each project, as well as a central board of review to which the local boards would report. Each local board would conduct hearings on a designated project and make recommendations regarding what payments the settlers should be obligated to pay, and how much relief they should receive based on circumstances beyond their control.

It was within this context that a review of the suitability of Bard lands for irrigation was conducted. On most projects, a committee of three, consisting of a representative of the Reclamation Service, a representative of the water users on the project, and a person connected neither with the Reclamation Service nor with the water users, was appointed to investigate problems. Because Yuma's original review board was dissolved for unspecified reasons, its reconstituted local review board was composed of just two representatives, both of whom were civil engineers and university professors who were judged to be impartial representatives. Elwood Mead, a professor of rural institutions at the University of California–Berkeley, was designated chair of the review board. His associate was Burton P. Fleming, a professor of civil engineering at Iowa State College. Both individuals were acceptable to the Bard water users and to the Reclamation Service in terms of their ability to be unbiased and objective concerning their investigation of the Yuma Project.[36]

After completing hearings and an inspection of the project, the board concluded:

> The moral responsibility for these settlers' plight . . . should not be ignored. They were told that this land was productive. The advertising pamphlets gave glowing

examples of large acreage returns. They were told that the land would be drained. The fact is, the land is not productive and it cannot be drained at any reasonable cost. The situation as we see it is that both the Reclamation Service and the settlers were deceived by appearances. Both are bound to lose something because of this, but the stronger party ought to stand by the weaker in preventing a disaster to either.[37]

This statement underscored the ineptitude of the Reclamation Service in connection with its handling of the settlement process at Bard. Mead's personal experience with the Yuma Project and the lessons learned there would serve him well in later years, for in 1924 he would be appointed commissioner of the newly reorganized agency known as the Bureau of Reclamation.

In an initial attempt to make reparations to settlers, early in 1916 a survey of the Bard district was arranged to determine the exact acreage impacted by seepage, alkali, or sand in order to ascertain the influence of soil conditions on the ability, or inability, of farmers to produce paying crops. The affected water users would then be temporarily relieved from payment of charges on land rendered nonproductive for causes beyond their control. An inspection team personally visited each of the farm units between late February and early March in 1916 and recommended the temporary elimination of 3,170 acres from the irrigable category.[38] This represented nearly one-half of the land encompassed by the Bard unit. Approximately one-third, or 55, of the 173 farm units were considered entirely nonirrigable. Most of these parcels were located adjacent to the levee (map 8.3). Only 32 farm units were classified as fully irrigable. They tended to be located to the north and west of the agricultural experiment farm. The majority of the remaining farm units were at least 50 percent irrigable. Most of the nonirrigable tracts, amounting to 2,800 acres, resulted from seepage, including 1,250 acres that also contained high levels of alkali. Another 170 acres were regarded as alkali-prone, but with no seepage noted, and 200 acres were classified as sandy.

When the word spread that relief for some settlers was under consideration, in mid-June 40 unit holders at Bard petitioned project manager L. M. Lawson to include an additional 1,000 acres scattered across their holdings in the lands earmarked for elimination. Even though Reclamation felt that their request was largely baseless, since some units simply were not being worked properly by their owners (hence the inability to yield paying crops in such cases was not due to soil or seepage conditions), another inspection was ordered for the district.[39] The second inspection was carried

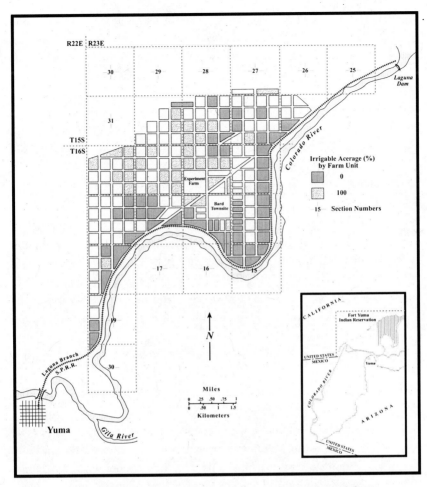

MAP 8.3. Nonirrigable versus fully irrigable farm units in the Bard district, March 1916. (Compiled from deductions made on account of seepage, salt, and sand. Inspection of units by S. P. Huss, February 22 to March 4, 1916. File 560-A6, Box 1104, National Archives, Washington, D.C., RG 115)

out in early July, at a time when conditions were significantly different from when the original inspection was made. Now the seepage lands looked better than before, and the sandy lands looked poorer.

The original survey, it turns out, was conducted at an unfavorable time, occurring soon after an unusually high January flood that caused a portion of the Reclamation levee to fail, inundating about 2,000 acres in the Bard district. As a result, the sandy lands, because of the silt deposits from the floodwaters, looked better in the spring, whereas the general seepage conditions because of the high river were worse than normal. And because of the unusually low river during the second inspection, some of the land eliminated earlier in the year due to seepage was now producing bumper crops, and the farmers on the seeped land were using practically no water to do so.[40] The July inspection, therefore, resulted in a revision of the original report, restoring 213 acres of seepage land, but eliminating an additional 715 acres of sandy land. Now approximately 3,670 acres were recommended for elimination, 2,590 of which were affected by seepage, 910 classified as sandy, and 170 acres evaluated as alkaline with no seepage.

The net result was that an additional 501 acres were recommended for elimination. The largest concentration of acreage removed from the irrigable category was still located adjacent to the levee, but nonirrigable land was scattered throughout the Bard district. It was not the intent, however, to permanently remove the majority of these lands from cultivation, for it was anticipated that more experience and adaptive farming methods on the part of the water users, and the completion of an adequate drainage system, would eventually restore much of the area to the irrigable category.[41]

Two years later, in March 1918, Lawson's successor, W. W. Schlecht, revised the earlier figures regarding the elimination of lands from the irrigable category. From Schlecht's standpoint there were four classes of land in the Bard unit: good, fair but sandy, fair but alkaline, and seeped.[42] Approximately one-half of the Bard unit, or 3,300 acres, consisted of good soils and did not as yet require drainage or were now protected by the partially completed drainage system. Fair soils constituted another 2,000 acres. Fair but sandy soils required an excessive amount of water and were lacking in plant food, but with proper farming combined with a recommended reduction in the cost of irrigation water could pay returns on crops. Fair but alkaline soils due to the high water table, if handled properly, could raise paying crops with relatively small amounts of water. Seeped lands, which totaled about 1,200 acres, were covered with water during the high stages of the Colorado River and therefore were not dependable in terms of crop yields.

Schlecht pointed out that the previous report completed under his predecessor eliminated from the irrigable class nearly all but the good soils, whereas Schlecht believed that the twelve-year graduated schedule of payments combined with adjustments in water charges would provide sufficient relief to settlers who were occupying farm units containing fair soils, assuming they were serious about bringing their farms under cultivation. On the other hand, the average farmer on seeped lands would not be successful, although some unit holders were turning to raising date palms on these lands. Accordingly, Schlecht recommended that farmers located on the seeped lands be given the option to relocate to farm units in the Yuma Valley. But only eight unit holders ultimately requested to be transferred to the Yuma Valley.[43]

Ironically, most unit holders with seepage problems remained on their farms with the unrealistic hope of eventually pursuing the lucrative enterprise of date production. The report of the first government soil survey team in the region inferred that date palms would thrive on seeped and alkali-prone lands, that is, under environmental conditions that would generally preclude the production of other crops.[44] And results from the local agricultural experiment station demonstrated that seeped land appeared to be ideal date-growing land. In 1900, Hall Hanlon, a longtime resident of the area whose farm was located next to the river, in fact, had several date-bearing trees standing from thirty to fifty feet tall.[45] Although promising, it took several years before dates would bear a profitable crop, and the majority of settlers simply were in no position financially to be able to explore the possibility of date culture.

Meanwhile, drainage work at Bard continued. After the failure of the completed drains during the river's flood stages in 1914, the Reclamation Service concluded that drains of a much larger capacity and at more frequent intervals would have to be constructed.[46] Early in 1915, tile drains were repaired and expanded in the southern and eastern portion of the Bard unit, lowering the water table by two feet.[47] It was found that tile drains would work efficiently if properly placed. But further drainage work in the Bard unit was now classified as supplemental construction, to be paid by the water users, as required by the adoption of the twenty-year extension act, and by mid-1915 drainage work came to a halt, pending a vote by the water users on the matter of repayment of drainage expenditures. It was nearly two years later, in April 1917, before the referendum was taken, with the majority of water users voting against paying for additional work of this nature.[48] According to one farm unit owner, additional drainage expenditures were refused because the experiment stations in the Yuma Valley and at Bard had long advised settlers that seeped land was ideal date land, that dates would do well with "their feet in the water and their head in the fire."[49] At Bard, seepage furnished the water and the summer sun provided the fire. And the same individual noted that the date orchard of the

University of Arizona Agricultural Experiment Station in 1917 earned a net profit of one thousand dollars per acre.

Although dates were well adapted to the climate and could potentially bring high returns, the expansion of acreage in dates was hindered by the scarcity and high price of young palms for planting. A decade later only about 30 acres had been planted to dates on the project, and most of the trees were too young to produce large crops.[50] Hence, a prospective date farmer would have to adopt a more immediate cash crop in the interim. Many farmers had by now discovered that, after digging small open-cut drainage ditches that would drain pools of standing water on seeped land, they were able to grow cotton, barley, and alfalfa on their farm units after all.[51] Then in 1920, another major flood on the Colorado River caused the large meander that washed up against the Bard unit to be cut off, thereby significantly lessening the seepage problem. This shift in the river also caused nearly ten square miles of the state of Arizona to be suddenly positioned on the west side of the river. After ten years of distress, the myriad problems experienced by Bard unit holders were gradually being overcome and farming on a paying basis was finally taking root.

Allotment Leases

The Indian unit adjacent to Bard suffered under a different set of problems, not the least of which was a general lack of interest among the Quechans in working their allotments. Seepage was not a significant concern here because the channel of the Colorado, even before its shift, was more distant from the Quechan allotments. While the Quechans awaited their allotments, many were employed by farmers next door in the Bard district, helping to clear and level farm units and to plant and harvest crops, whereas others were employed as day laborers for the Reclamation Service.[52] The Quechans also were given a 20-acre temporary planting place that was provided with water by the Reclamation Service.

The Indians who began immediately to farm their allotments were evidently those who had gained some experience by working for the Bard settlers. But little in the way of improvements on Indian allotments occurred for several reasons, the most important being that the majority of Quechans simply were not eager to begin farming in the white man's way. This was partly due to the fact that the Indian Office was negligent about providing the Quechans with the necessary advice and equipment to farm, including teams of horses, which were needed to begin the heavy work of clearing, leveling, and planting the land. And the 1912 flood on the Colorado River made it possible for the Indians, for the first time in five years, to clear patches of overflowed land not protected by the levee and farm them as in former times. The Quechans subsequently were able to harvest a surprisingly large crop of beans, pumpkins, melons, and corn in 1912, paradoxically causing the superintendent of the Fort Yuma Indian Agency to express his concern that the Quechans' good fortune might deter them from taking an interest in their allotments.[53]

By the end of 1913, only 136 acres scattered among the Quechan holdings were irrigated, and only 5 acres of the 160-acre Indian School farm were under cultivation. This poor showing reportedly was partly due to the Quechans' distrust of the government's intentions toward them. Although all of the 10-acre allotments had been assigned by April 1912, the secretary of the interior did not formally approve them until nearly a year and a half later, in September 1913, and the Reclamation Service had not as yet completed its system of laterals to deliver water to all of the allotted parcels. As of late 1913, Indian allottees located in the western part of the Reservation Division were still awaiting laterals to serve their allotments (map 8.4). In fact, there was so little progress being made on the allotments that a smallpox epidemic, which swept the reservation during the winter of 1913–1914, was seen as a positive event because the Quechans were quarantined and issued rations to work on their allotments: "Some work was done that otherwise would not and added materially to the total for the year," reported the superintendent.[54] In other words, some allotments that otherwise would have stood idle began to be cultivated. The demand for Quechan labor in the area, however, remained strong, and after the quarantine was lifted there was no compelling need for the Indians to continue to work their allotments.

Failure to finish the system of laterals in the Indian unit before the allotment process was completed contrasted sharply with settlement policy in the Bard district, where the water delivery system was ready when the land was opened to settlers, and this distinction exhibited a clear bias by the Reclamation Service in favor of the Bard unit of the project. In fact, after the Bard unit began to be settled, two substantial fences were built by the Reclamation Service "for the purpose of separating the Indians' from the Settlers' lands."[55] This "iron curtain" was also a symbolic boundary marking differences in construction policy on each side of the fence. Not only did the mind-set concerning the extension of laterals change when this boundary was crossed, but so did the materials used in construction. For example, concrete, a more expensive and permanent material, was adopted for the construction of

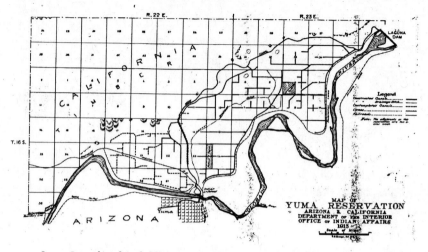

MAP 8.4. Completed (*solid lines*) and contemplated (*dashed lines*) irrigation ditches in the Bard and Indian units, 1913. (Reproduced from "Map of Yuma Reservation, 1913," courtesy of the Perley M. Lewis Manuscript Collection, Department of Archives and Manuscripts, Arizona State University)

irrigation structures in the Bard unit, whereas more temporary and less costly wood structures were being built in the Indian unit. This form of discrimination was not unique to the Quechans. Indians, being a minority group, commonly were treated differently by white-dominated institutions.[56]

The reasons advanced by the Reclamation Service for these differences was that, in the Indian unit, the service felt that construction should proceed according to the amount of agricultural activity that might be expected from the Indians.[57] Since so little use had been made of the completed laterals, additional construction did not seem justified, the service reasoned. And the Reclamation Service attributed the Quechans' lack of interest in their allotments to the Indian Office, which had provided no material assistance to the Indians in bringing their land under cultivation. The service was therefore opposed to spending more money on the extension of the laterals until some tangible assistance at least was furnished the Indians. The remark was made at the time that "certainly the Reclamation Service should not be expected to advance the money for the completion of the irrigation system if the Indian Department is not willing to even trust the Indian with sufficient implements to till the soil."[58] Additionally, the Indian Office was already two years in arrears in repayments to the Reclamation fund for the irrigation works that had been completed. The Indian Office, on the other hand, countered by claiming that it should not have to pay for an irrigation system that had not as yet been completed.[59]

In order to reconcile the differences between the two federal agencies, Cato Sells, commissioner of Indian affairs, and A. P. Davis, who had recently succeeded Newell as director of the Reclamation Service, agreed to arbitration. A board of arbitration was appointed in late January 1915 charged with the task of offering suggestions that would "lead to harmonious and equitable relations in the future operations at Yuma between the two bureaus."[60] Within a few days the board had completed its task. It was found that although water had been delivered to each 40-acre tract on lands occupied by Bard settlers, it had mainly been delivered at the heads of the completed laterals on lands occupied by the Indians, resulting in increased costs to the Indian Office in getting water to the Indian allotments. It was recommended, therefore, that operation and maintenance charges be lowered for the Quechan allotments because of this inequity.

On the matter of concrete versus wooden structures, the board urged conformity in construction between the Bard and Indian units, unless soil conditions at any particular structure site prevented the use of concrete construction. Regarding the completion of the remaining laterals in the Indian unit, the Reclamation Service had already committed, prior to arbitration, to a target date of mid-July 1915, and the Indian Office had registered its satisfaction with this completion date.[61]

The Reclamation Service had reversed its policy concerning the matter of further extending the laterals on the Indian unit because local officials of the Indian Office had requested that the lateral system be finished "in order that some Indians who desire could farm their allotments, and that other allotments could be leased to white men who were seeking leases on the reservation." Chief engineer W. M. Reed of the Reclamation Service rationalized this change in policy by stating, "We cannot expect the Indians to develope [sic] their claims, and we cannot expect white men to lease these various allotments until there is an assurance that when they have done the preliminary work, which is quite expensive, they will receive water for irrigation."[62]

The underlying assumption now was that the quickest and best way to get the Indian unit under cultivation was to allow the Quechans' allotments to be leased to non-Indians, and work by the Reclamation Service was immediately begun to complete the laterals in the Indian unit. Although the arbitration board had noted that because of the Quechans' prevailing distrust of the government the installation of irrigation structures prior to the preparation of the land on the Indian unit would give the Indians confidence that they would be able to get water after their allotments were prepared, by now there was an awareness among officials of both agencies that the Quechans themselves generally would not be the ones to prepare the land for cultivation.

It was evident as far back as the 1890s that assigning lands in severalty to the Indians would not necessarily turn them into farmers, and it was obvious among Reclamation Service and Indian Office officials even before the irrigation works on the Yuma Project were completed that the Quechans would not readily adopt irrigated farming on their allotments. Nevertheless, newly appointed Superintendent Loson Odle, upon his arrival at the Fort Yuma Indian Reservation in 1911, maintained that the Indians should be compelled to farm rather than lease their lands. And the following year he remarked that "a great awakening has come to the Yuma Indian within the past two years" by watching and helping the white settlers.[63]

The superintendent's outlook toward leasing would quickly change, however. In 1914, after Indian trust patents for the allotments had been issued, he realized that the Quechans had little interest in farming their allotments, and that his own agency had even less interest in helping them do so. Turning the Quechan allotments into a successful farming enterprise for the allottees would require more abundant aid than the Indian Office was willing to provide at the time. Odle therefore contacted Cato Sells, commissioner of Indian affairs, for permission to lease allotments to non-Indians. "It will prove advantageous to the Indians in general," he wrote, "to have all lands belonging to women and children, with exceptional cases, arbitrarily leased for a period of five years or less for improvements including clearing, leveling and placing in crops, fencing, etc."[64]

By arbitrary leasing, Odle meant that he as agent would arrange the leases; the Indian allottees would have no say in the matter.[65] The leases generally ran for a term of five years, after which the leased allotments were to be left in alfalfa, with the prospect that they would then be farmed by the Indians. It was Sells's predecessor, Robert Valentine, who had extended the limit of agricultural leases to five years and encouraged rental by white farmers. Under Sells's administration the Indian Office was expanding its leasing program, and he therefore approved Odle's plan, but Odle was reminded that the ultimate aim of the allotment program was to induce the Indians to farm their own land.

The General Allotment Act, when approved in 1887, contained no provisions for leasing because it was assumed that Indians could not learn farming if someone else worked their allotments. Even before President Arthur's executive order created the Quechan reservation in 1894, and long before Congress approved the allotment of the Quechan reservation in 1904, much of the previously allotted land on other reservations remained unfarmed by the Indians. It was non-Indian farmers and ranchers, therefore, who increasingly pressured Congress to open Indian lands for white use.[66] Hence, in 1891, Congress gave those Indians who for reasons of age or disability could not work their land the right to rent it with the approval of their agent. The following year Congress added a new elastic category—"inability"—to the list of reasons that justified leasing.[67] The rationale now was that rental income would benefit the Indians who were too poor to farm, and if the Indians had a supplemental source of income from their leases, they could gradually purchase the tools necessary to eventually work their allotments. Also, if non-Indian farmers made initial improvements to the allotted land, Indians would be more likely to succeed when they did begin to farm, because their land would have already been prepared and the Indians would have been able to learn to farm by example.

This line of reasoning set the stage for the hundreds of improvement leases that were approved by Odle on the Indian unit of the Yuma Project. By mid-1915, when the irrigation system was nearly complete as promised, only about 1,000 acres in the Indian unit had been improved, while the remainder was still "awaiting the Indian's ax, plow, etc." This situation quickly changed. Since the cost of clearing and leveling the allotments averaged $75 per acre, most Quechans simply could not compete with better-capitalized Anglo farmers, and their allotments were mostly put under cultivation by leasing them, for a five-year term, to non-Indian tenants. The area under cultivation rapidly increased to more than 8,000 acres by 1920 under an improvement lease program, 6,200 acres of which were leased, while the Quechans farmed 1,820 acres.[68]

Much of the demand for Indian leases came from Bard unit holders, who found lands in the Indian unit better suited to cultivation.[69] Essentially, Odle had approved the leasing of three quarters of all allotments, 620 out of the original 819. This dramatic increase in the number of leases on the Quechan reservation was paralleled elsewhere as the prosperity associated with World War I invigorated the general leasing program. Once the United States entered the war, Sells launched a patriotic campaign to increase food output from Indian trust lands, and, by 1920, 4.5 million acres of trust land were under lease.[70]

When the official notice of approval of allotments was received at Yuma on January 1, 1914, plans for leasing Quechan lands were made at once. The leasing program was looked upon by the Quechans with suspicion; they viewed it as another scheme whereby even more of their lands would be taken by non-Indian settlers.[71] The Quechans were thoroughly disillusioned; they could not understand why agent Odle had the right to lease their lands without their consent, and they felt the government had seriously betrayed them by not furnishing them with farm implements and other assistance.[72] The number of leases was relatively small in the beginning because Odle believed then that the Quechans would farm their own land; but by mid-1915, when the entire system of laterals was fully operational on the reservation, almost 90 percent of the Indian unit had not as yet been cleared and leveled, even though most of the Quechans had established residence on their allotments, living in houses built of poles and filled with mud.[73]

In January 1916, a break in the Reclamation levee inundated nearly the entire Indian unit with four to ten feet of water, destroying irrigation ditches and the Quechans' dwellings. Before farming could be expanded in the Indian unit, the irrigation ditches had to be reconstructed and repaired, most of the work being done by the Indians themselves. Nevertheless, by mid-1916, 120 allotments were leased, and the number of leases progressively increased to 300, 500, and 572 in each successive year, achieving a high of 620 allotments by mid-1920. This trend ran counter to Odle's statements in 1916 when he maintained that no lands would be leased where the Indians could possibly farm them for a profit and affirmed that less land, not more, would be leased each year.[74] But the leasing system was the inevitable result of the allotment system, which in turn led to the continual decrease in acreage farmed by the Indians.[75]

Odle rationalized the rapidly growing improvement lease program by arguing that it was impractical for the Indians to do all of the preliminary work for cultivation, and that at the expiration of the leases the land would be left in alfalfa. "At that time," he wrote, "many of the Indians will farm all of their lands." Odle also pointed out that because of the leases, white families were living among the Indians, and hence a "better feeling is developing between the races." But by 1923 a large share of the improvement leases had expired and, in most instances, were renewed for cash rental. Odle now related that "the lessees, as a class of men, have been a very fine example to the Indians in their farming operations," and that the renewal of leases kept the land from lying idle while providing the Indians funds to work the lands they retained.[76]

Eventually, the 160 acres reserved for the Powell townsite and the north 80 acres of the school farm were made available for allotments—and leasing—for those Indians whose original allotments turned out to be of little agricultural value. During Odle's long tenure as Indian agent at Fort Yuma, which ended in 1925, little was done to promote the Quechans' own farming efforts. Nevertheless, even as the end of his term approached, he asserted, "When they [the Indians] have mastered intensive cultivation . . . they will be independent."[77]

Leasing, however, subverted the original purpose of allotment. Indian landlords had no reason to farm or become self-sufficient. The leasing plan not only was a contradiction of the official objective of the allotment policy but also denied the Quechans the full economic benefit of the only economic resource, other than their own labor, that they possessed.[78] From 1913 to 1920, the Quechans averaged $150 per capita annually as day laborers.[79] The reasoning was that if everything went as planned, the primary cost of developing the land would be borne by white farmers through the improvement leases, and the Quechans, if they could hold out for a few years until the land was improved, could then farm their own allotments. Leasing for improvement, however, was eventually replaced by leasing for cash, with the leases extended to ten years, and this, not allotment farming by the Quechans themselves, became the basic source of income from the use of their land after 1914.[80]

The cash rentals, however, turned out to be a pittance, netting the Quechans between $30 and $70 per year for each 10-acre tract. Under the terms of the Quechan leases, the allottees were expected to pay operation and maintenance charges on the irrigation water delivered to their allotments. The charges were deducted from whatever lease money was received by the agency, on the grounds that because water charges arose from improvements made on private lands, the Indian allottee and not the lessee or the tribe as a whole should bear the costs.[81] And the lessees, who had no vested interest in the land, but of course were interested in making as large a profit as possible, induced serious environmental damage. Repeated overirrigation eventually led to the nonirrigability of sizable tracts within the Indian unit due to increased salinization.

Through its unsuccessful attempts to turn the Quechans into self-sufficient farmers, the government forced the Quechans into land leasers and menial wage laborers, who were preoccupied with day-to-day existence rather than long-term economic growth.[82] Assimilation of the Indians into white society had long since been abandoned as the central concern of policy makers. By 1920 the idea that the Indians would control their lands and rise to "civilization" with the aid of the federal government was now rejected. In the final analysis, permitting the leases was an expedient move by the government, enabling Indian agents to maximize production on Native lands while saving the government complete embarrassment over the failure of both the Reclamation Service and the Office of Indian Affairs to successfully implement their respective programs. At the same time, it also opened even more reservation land to non-Indian farmers. According to Frederick Hoxie, the policies of the Indian Office ultimately "placed the Indians on the outskirts of American life and promised them a limited future as junior partners in the national enterprise."[83] Yet the Quechans were relatively fortunate compared to other tribes because title to their allotted land remained in government trust, effectively keeping their lands from passing permanently into the hands of non-Indians.

Yuma Valley Farmers' Frustrations

Developments in the Yuma Valley did not go smoothly either, even after the siphon was completed and adequate supplies of water began to be delivered to valley farmers. In 1912, just as the siphon was being placed into service, the high water and swift current of the Colorado River destroyed three miles of the Reclamation levee protecting the Yuma Valley. By now it had become apparent to the service that taming the Colorado would be much more difficult than originally anticipated. Meanwhile, after the siphon was completed and the levees repaired, the Reclamation Service was prepared to deliver far more water than the farmers were ready to use.

In 1913, the distribution system covered about 45,000 acres below Yuma, but only about 14,000 acres were actually receiving water.[84] Part of the problem was that land that had been cleared and leveled in the past in anticipation of the arrival of water had grown up with brush and had to be cleared again. The valley lands also were slow to develop because of the prevailing difficulty among farmers at the time to borrow money at reasonable rates of interest to help cover the cost of preparation and planting. Additionally, the extremely poor road conditions in the valley presented a problem for Yuma

Valley farmers in transporting their crops, mostly alfalfa, to nearby markets. Farmers, who had been receiving water in the past, had depended largely on alfalfa production because the remoteness of their lands from major markets precluded intensive cultivation, and because alfalfa generally could be profitably marketed in the vicinity. The long haul to Yuma, however, over bad roads from newly cultivated land in the lower reaches of the valley made the task of transporting the alfalfa crop exceedingly difficult. Under the circumstances, in 1913 supervising engineer Hill advised director Newell that it would be unrealistic to expect farmers to put their land in cultivation at a more rapid rate than 10,000 acres per year.[85]

The Reclamation Service, however, believed that two of these obstacles— the river's attacks on the levee and inadequate transportation facilities— could simultaneously be solved by the construction of a railroad. The service now recognized that the Colorado River changed location quickly and often in the vicinity of Yuma, and that the levees below Yuma would require substantial maintenance efforts to protect project lands from flood damage. Brush protective works had already proved ineffective; the only practical way to protect the levee from attacks by the river involved rock revetment, but this would be possible only by use of a railroad. Consequently, project engineer Sellew contacted the Southern Pacific Railroad regarding the possible construction of a rail line in the Yuma Valley south of Yuma to facilitate the work of the service. Sellew pointed out that not only would the railroad help maintain the levee, but it would also be of considerable assistance to the Reclamation Service for transporting materials that would be needed for the completion of the water distribution system in the valley, while also hastening the development of the community by providing a means of transportation for farmers to get their crops to market at Yuma.[86]

Since the Reclamation Service had contracted with the Southern Pacific in 1907 to build the "Laguna Branch" railroad along the top of the Reservation Division levee on the California side to provide rail service for the construction of Laguna Dam, Sellew assumed that the firm might be interested in doing the same on the Yuma Valley levee. For the first few years the Laguna Branch was profitable to the Southern Pacific because of the construction traffic. However, after the dam was completed and the branch line was forced to rely principally on passenger and settler-generated freight traffic, it proved unprofitable. Undoubtedly because of this experience, in a letter dated March 30, 1914, representatives of the Southern Pacific declined the offer.[87]

By that time, however, a board of consulting engineers had submitted a report to director Newell supporting Sellew's view regarding the necessity of a railroad on the levee, even if the Reclamation Service had to build it, so that attacks from the river could be resisted by the rapid placement of rock. Hence, one month after the Southern Pacific's rebuff, on May 1, 1914, the Reclamation Service drove the first spike of the Yuma Valley Railroad, and by the end of February the following year the twenty-three-and-a-half-mile standard-gauge line was completed. Although the service built the railroad primarily to protect the levee, it also hoped to reap an ancillary benefit—the more rapid development of the Yuma Valley by facilitating the transport of goods produced there. In fact, just as construction began on the rail line, a group of citizens petitioned for a spur line on the railroad. This was followed by many other requests for special spurs, sidings, and extensions of the line. Although the Reclamation Service hoped the railroad would promote development in the Yuma Valley, it never extended the railroad into the valley after completion of the main track. It did, however, permit spur tracks to be built, at others' expense, in order to increase use of the main line.[88]

Passenger service on the line began on January 12, 1915, for the opening sale of lots in the newly established town of Gadsden. At first, service consisted of six round-trips each week, but was reduced to three trips a week after the January 1916 flood stranded the motorcar on the line. Passenger as well as freight service on the line consistently lost money. Hence, in early 1923, passenger service was discontinued, followed in 1929 by the termination of freight service.[89] But during the 1920s, the Yuma Valley Railroad had an indirect influence on the improvement of transportation in the valley, for it hauled major shipments of sand and gravel for road construction, eventually making truck hauling on improved roads a practical alternative to the use of the railroad.

If the railroad had been built strictly as a commercial enterprise, it no doubt would have run through the center of the Yuma Valley, connecting Yuma with Somerton and Gadsden, not on top of the levee along the valley's western edge. Nevertheless, the rail line, in connection with the economic boom stimulated by World War I, did impact development in the Yuma Valley. And as a means of transporting rock to be used for the protection of the levee, the railroad was considered an unqualified success. Its operation was especially valuable in the summer of 1921 when the Colorado River reached its highest flood stage on record and the levee was breached just north of Gadsden, submerging nearly seventeen hundred acres.[90] It continued to serve this purpose until the mid-1930s, when the completion of Hoover Dam far upstream, with the huge storage capacity of Lake Mead behind it, brought the lower Colorado River under complete control, after which the rail line was no longer needed for rock revetment work. After the last of the rock trains was operated in June 1934, the Somerton Chamber of Commerce requested that the railroad be made available for shipments of winter vegetables, particularly lettuce. Anticipating that such service would encourage development of winter vegetable farming, freight shipments were resumed by the Reclamation Service. By 1940 the major freight carried on the line included hay, grapefruit, cantaloupe, lettuce, miscellaneous other farm produce, and livestock.[91]

But during the first few years after water began to be delivered to Yuma Valley settlers, the most compelling concern among water users involved the construction charge that they would ultimately have to bear when the public notice for the valley was finally issued by the secretary of the interior. Until then they were receiving water on a rental basis at a bargain rate of fifty cents per acre-foot from the Reclamation Service without having to repay the cost of construction, but the impending public notice would require that settlers begin repaying the government under the twenty-year extension act for their share of the irrigation works.

As with most of the early projects, the Reclamation Service met with substantial cost overruns in completing the Yuma Project, with the original estimate of less than $3 million ballooning into a final cost of $12 million. Nearly all early project estimates were made between 1902 and 1906, an inflationary period exaggerated in the West by unique regional conditions, while Reclamation personnel relied on existing prices to calculate future expenditures for labor and material. Also, water users often sought construction of drainage systems and other supplemental works not included in original estimates.[92] In terms of percentage increase over the original estimate, Yuma was the third highest among early Reclamation projects.[93] This meant a potential increase in cost to water users from the initial estimate of approximately $35 per acre to perhaps $100 per acre, and unit holders in the Yuma Valley considered it unfair that they should have to shoulder the increased cost. Hence, in August 1915, while Mead and Fleming were conducting interviews with Bard unit holders over the irrigability of Bard lands, water users in the Yuma Valley requested hearings regarding the anticipated construction charges that would be levied against them. Even though the board of directors of the Yuma County Water Users Association did not approve of participation in any such hearings by its members, a petition nevertheless was circulated among the general membership that read:

> The undersigned, water users and landowners under the Yuma Project, Arizona side, heartily approve of your selection as members of a Board of Review to investigate conditions existing upon the Yuma Project and to make recommendations for such action as will, in your judgment, do equity, promote the prosperity of the water users, and insure the success of the Project.
>
> While we realize that neither the Government nor the water users will be in any way bound by the Board's action or report, we feel that your recommendations will go far in solving the difficulties that are met with in this project.[94]

Since the petition was signed by more than a majority of water users, representing more than half of the privately owned acreage in the Yuma Valley, the board accepted the invitation, and hearings were scheduled in mid-October at Yuma, Gadsden, and Somerton. During the hearings, farm unit holders in the Yuma Valley argued that the project cost to be repaid by them had been fixed in May 1904 at between $35 and $40 per acre, in accordance with the estimate included in chief engineer Newell's letter to them at the time, countersigned by the secretary of the interior, outlining details of the project. More specifically, later in that month, when stock subscriptions in the Yuma County Water Users Association were signed whereby unit holders placed a lien on their land for the receipt of water, the assumption at the time was that their acreage assessment would be approximately $35. The review board conceded that it was not qualified to determine whether the circumstances under which land owners placed a lien on their lands constituted a contract that was legally binding on the government, but when a formal contract was finally prepared and approved by the Yuma County Water Users Association two years later, the water users did agree to "guarantee the payment of that part of the cost of the irrigation works which shall be apportioned by the Secretary of the Interior to its share holders."[95] This contract, therefore, from the standpoint of the board, appeared to ensure repayment of the actual cost of construction, whatever that figure might be.

Assuming this to be the case, Mead and Fleming then endeavored to formulate a figure that would be fair to Yuma Valley landowners in light of the extraordinary increase in the cost of construction. They noted that one of the primary areas responsible for increased costs involved riverfront protection, because the obstacles to be overcome in diverting and controlling the Colorado River simply were not foreseen when the initial estimate was made. The board therefore recommended that all expenditures on levees, railroads, and riverfront protection be eliminated from the repayment charges, and that these be paid for by appropriations similar to those made for river and harbor works elsewhere.[96]

Mead and Fleming also recommended the abandonment by the Reclamation Service of the Gila unit of the project, an area encompassing 19,000 acres capable of being irrigated and where nearly $600,000 had already been spent on canals and riverfront protection but where only 900 acres were being irrigated. The North Gila Valley suffered from severe isolation due to lack of transportation facilities and relentless problems involving control of the Gila River, whereas the South Gila Valley, in addition to river-control problems, was automatically eliminated from the project when it was left without a source of water supply for irrigation after the original location of the siphon was changed from the Gila River to the Colorado River below the Gila confluence. Considering the fact that the levee system built to protect the Gila unit already had largely been washed away, it seemed futile to continue to try to develop this unit, and to do so would be throwing good money after bad. If these and certain other modifications suggested by the board were made, the project cost for the Yuma Valley lands would be so close to those fixed for the Bard lands that it was recommended the cost be the same, that is, $55 per acre.[97]

The secretary of the interior, however, only partially concurred with the board's recommendations. On April 6, 1917, when the public notice was issued for the Arizona lands, it included only those in the Yuma Valley, excluding the Gila unit as recommended, but the construction charge was set at $75 per acre, not the suggested $55 per acre. Although $75 was just over half of the final actual per acre cost, which turned out to be $142.80 (second highest among all early projects), the final figure was met with displeasure in the Yuma Valley, and the Yuma County Water Users Association promptly filed suit against the government, citing project manager Schlecht as the defendant. On the other hand, the construction charge levied on water users apparently did not seem excessive to potential future entrymen on the project. In the summer of 1918, when 16 farm units constituting the remaining unentered public land in the Yuma Valley were thrown open for settlement, more than 1,000 letters of inquiry and a total of 724 applications were received for the newly offered farm units. These units, all of which carried the $75-per-acre construction charge, were allotted at a public drawing held at Yuma in December 1918.[98]

Nevertheless, the Yuma County Water Users Association suit was brought to trial in the U.S. District Court in Tucson in late April 1919. The plaintiffs alleged that the water-right application required to be filed under the public notice of April 6, 1917, was an unconscionable contract because it carried an agreement to pay a construction charge of $75 per acre instead of the $35 to $40 an acre consistent with the original estimate. They argued that had subscribers to the Yuma County Water Users Association known that the cost would exceed the estimated amount, they would not have signed such stock subscriptions, particularly since a private venture, the Colorado Delta Canal Company, attempting to enter the field at the time, provided an alternative irrigation system for the Yuma Valley at a lower cost.[99] Additionally, the Reclamation Service, as recommended by the board of review, had officially eliminated the Gila Valley unit from the project, and in 1918 the operation and maintenance of the extant irrigation system were turned over to the North Gila Valley Irrigation District for a period of twenty years. According to the plaintiffs, since the original contract contemplated the construction of irrigation works for the Gila Valley unit, and since the public notice for the Arizona side of the project failed to include this unit, the project was incomplete and payment of the construction charge, as prescribed by the public notice, could not legally be imposed.[100]

The trial was held in late April 1919. Testimony was given by A. P. Davis, director of the Reclamation Service; J. B. Lippincott, the original supervising engineer of the Yuma Project; Francis Sellew, who served as project engineer from 1906 until 1915; Oscar Bondesson, the first presiding officer of the Yuma County Water Users Association; and others during the six-day trial. Much of the testimony focused on the causes of escalating costs and whether the project was indeed finished. In addition to levee maintenance, one of the leading causes contributing to the increased costs was that the scope of the work provided by the Reclamation Service changed over time. Initially, it was generally assumed that the service would be responsible for constructing dams, reservoirs, and main-line canals, leaving the farmers the task of building the lateral distribution system. This was consistent with the practices of pioneer irrigators in the West, who, individually or cooperatively, worked together to take water out of the smaller streams. The original plan, then, was for the government to provide the "artificial streams" from which water might be taken by settlers.

It was soon realized, however, that the construction of irrigation works did not necessarily engender the expected collaboration among settlers. In former days, prior to federal involvement in irrigation, pioneers had only nature to fall back on while attempting to make the best out of sometimes unmerciful conditions, whereas those settling Reclamation project lands felt that they had recourse to the government if conditions proved too difficult. Hence, to bring irrigable lands under cultivation in the era of government intervention required the provision of a complete irrigation system by which the water would be taken to within a mile or so of each farm. The miles of small canals and laterals, and additional small structures such as flumes, culverts, bridges, and turnouts, added significantly to the original estimated cost of construction.

A large amount of the work on the Yuma Project also was done by force account, that is, it was not contracted out, resulting in higher costs for labor and materials. But project officers argued that accomplishing the work by force account enabled them to give employment to settlers, which would help tide them over during the early years of home building, and that this justified some increase in cost. Additionally, substantial increases in the cost of labor and material occurred in the West when the heaviest construction on most of the early projects was carried out.[101]

The government conceded that the elimination of the Gila Valley increased the cost per landowner on the project, but it was found impractical to control the Gila River within acceptable limits of expense, and locating the siphon crossing on the Gila River would have posed a grave hazard to the water supply of the Yuma Valley if the siphon were destroyed when the Gila River was in flood.[102] Regarding the issue of whether the project was actually finished because of changes in the original plan, the government argued that the current landowners of the community were able to carry out the business for which the project was designed, and it therefore was considered complete from the standpoint of the Reclamation Service.

In fact, in 1918 the Yuma Project had a gross yield per acre of more than $113, the highest of all early projects with the exception of the Okanogan Project in Washington.[103] In the end, the water users had a weak case, and, on February 18, 1920, judgment was rendered in favor of the government, meaning that the construction charge of $75 per acre set forth in the April 1917 public notice would stand.[104] Unlike Bard unit holders who had to fight for more liberal repayment plans, Yuma Valley water users could immediately take advantage of the subsequent liberalization of earlier repayment schemes, reimbursing the government on a graduated basis over a twenty-year period. And much of their land had by now already been prepared and planted, giving Yuma Valley farmers a significant edge before having to begin repayment. Given the value of crop production in the valley, and considering the benefit of being able to delay their repayment for several years after receiving water from the Reclamation Service, the $75 per acre charge handed down by the court did not appear to be an excessive burden for Yuma Valley water users to bear.[105]

For the first ten years or so after completion, nothing seemed to go right on the Yuma Project. The Bard unit holders experienced the same difficulties as their counterparts on other projects, having to begin repayment of construction charges before they could even glean a profit from their land. Compounding their problems were the environmental consequences resulting from the lack of drainage facilities, making portions of the Bard unit nonirrigable during high flood stages of the Colorado River.

On the Indian unit, contrary to the objectives of the General Allotment Act, the Quechans were forced to lease their holdings because so little was done by the Indian Office to help them bring their farm units under cultivation, and because of their general lack of interest in adopting irrigated farming. The five-year improvement leases that were originally permitted turned into ten-year cash rentals that did nothing to foster economic independence among the Quechans. And further complications arose resulting from the conflicts that developed between the Reclamation Service and the Indian Office over irrigation matters in the Indian unit.

The primary source of discontent in the Yuma Valley after the siphon was finally placed into service was the perceived excessive charges that would be assessed water users as a result of the exorbitant cost increases related to the completion of the engineering works of the Yuma Project. Meanwhile, the areas devoted to cultivation in the Yuma Valley expanded rather slowly because of inadequate marketing and transportation facilities. Given the naïveté of the Reclamation Service, combined with that of potential irrigators, and the range of unanticipated difficulties that the service and the settlers had to deal with, it is a wonder that successful irrigation farming ever became established on Yuma Project lands.

9

Reclamation and Retrospect

In many respects, the Yuma Project is a microcosm of the multifaceted difficulties found in various combinations on different federal irrigation projects. But in spite of the disappointments and conflicts connected with its early years of settlement and development, Yuma, because of its practically year-round growing season, also stood out among the early projects. In 1920, ten years after Colorado River water was first turned onto the land by the Reclamation Service at Bard, the Yuma Project was heralded as one of the most productive among the original twenty-one federal reclamation projects.[1] Yuma, it seemed, was now on its way to becoming one of the more prosperous agricultural districts in the country, even though initial prosperity largely revolved around the production of alfalfa and cotton.

A complement to Yuma's good fortune was congressional approval in January 1917 for the development of the Yuma Auxiliary Mesa Project. Because of its nearly frostless climate and good drainage, the Yuma Mesa was well suited to citrus production. Elwood Mead, in 1915, while serving on the local board of review at Yuma, declared that the redeeming feature of the Yuma Project would be the Mesa unit, because it was the only area that was not menaced by floods, seepage, or alkali.[2] It would take many years before the Yuma Mesa would come into full production, however, because the pumping plant that lifted water to the mesa from the valley below was not placed into operation until May 1922, and because of the substantial time lag between planting and harvesting citrus.[3]

Settlement and Agricultural Development

Meanwhile, alfalfa initially became the leading crop at Yuma as well as on nearly all project lands because it could most easily be grown by amateur farmers while they familiarized themselves with irrigation methods, and also because there was a ready market for this crop during the years that Reclamation work was in progress. In spite of the fact that climate conditions and the alluvial soils of the Yuma Project were more conducive to the

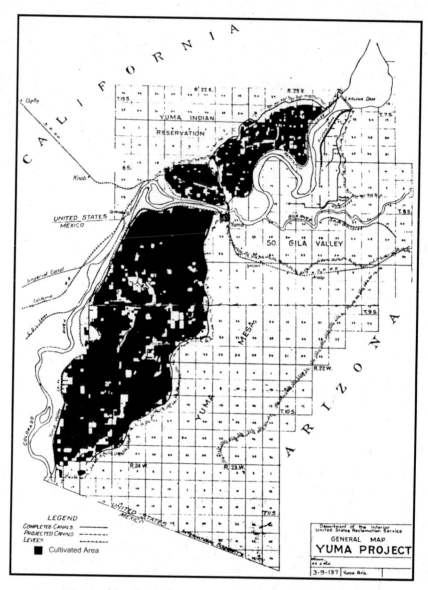

MAP 9.1. Irrigated farmland on the Yuma Project, 1920. (Reproduced from "Cultivated Area, 1920," in "Yuma Project Annual Project History and O. and M. Report, 1920")

development of diversified truck farming than that found on most other projects, this type of farming was limited in the formative years by small local markets, and by farmers who had little marketing experience, as well as by the extremely high transportation costs to distant markets that prevailed at the time. The intensive cultivation of winter vegetables, melons, dates, and citrus that characterizes the region today awaited the adoption of refrigerated trucking and the expansion of a more sophisticated nationwide road network that permitted travel time to distant markets to grow shorter. Even as late as 1927, date culture, the activity that from the beginning held great promise on the Yuma Project, was considered still in its infancy, with only a few trees in bearing.[4] But Yuma also lies in the shadow of the Salt River valley to the east and the Imperial and Coachella valleys to the west, and its relative isolation and comparative small size never allowed truly intensive farming to gain prominence until relatively recent times.

Although alfalfa hay became the staple of Yuma Project agriculture, it was alfalfa seed, which commanded a high price in the Imperial Valley and other alfalfa-growing districts, that quickly became the leading cash crop on the project. In 1913, 75 percent of the entire irrigated area on the Yuma Project was planted to alfalfa, yet the average crop value per acre was relatively high at $36.50, compared to an average on all projects of $24.50.[5] This figure, which placed Yuma second among all projects in per-acre output, was credited to the production of alfalfa seed, the yields of which were remarkably large because the region's long and severely hot summers were conducive to prolific seed yields. The common practice was to cut one or two crops of hay, then a crop of alfalfa seed, followed by three or four more crops of hay in one season.[6] Over time, alfalfa also had the beneficial effect of adding humus to the soil through stem, leaf, and root decay.

Cereal crops generally ranked second in acreage on early project lands, including Yuma. Although cereal grains contributed much less than alfalfa to the overall value of production, they did serve as a nurse crop for young alfalfa. The production of grain, grown as a nurse crop, or as an initial cash crop after breaking in the land, was a common practice in the early development of an irrigated farm.[7]

In 1914, there were 698 farms counted on the Yuma Project, with 25,200 acres under crop.[8] By the end of the decade, more than 53,000 acres were being irrigated on 1,225 farms, and the total crop value amounted to approximately $20.5 million.[9] Nearly the entire project was now under cultivation (map 9.1). In just a few years the acreage under irrigation more than doubled, while the number of farms on the project nearly doubled. Somerton, an agricultural service center platted in the heart of the Yuma Valley in February 1909, grew from a town of about 125 residents in 1916 to one of 1,200 in 1920. Much of this development during the latter half of the decade was related to a shift from almost total reliance on alfalfa production to cotton.

A downturn in the hay market caused by overproduction in the Salt River, Imperial, and Yuma valleys led many Yuma irrigators to turn to cotton as a profitable substitute. Cotton used less water and was less affected by high concentrations of alkali salts in the soil, and the crop also could withstand higher transportation costs to market. By 1916, the returns received from cotton surpassed those of alfalfa seed by $70,000, and in the following year the value of cotton produced on Yuma Project lands exceeded that of alfalfa by well over $1 million. In 1918, nearly 64 percent of the entire irrigated area of the project was planted to cotton, a figure that increased to 69 percent by 1920. Yuma, by this time, had gained the reputation of producing the highest-grade short staple cotton found on the market (fig. 9.1).[10]

Much of the expansion associated with cotton production occurred in conjunction with the growth of irrigated tracts in the Yuma Valley. At the end of 1915, the Reclamation Service was prepared to irrigate more than 50,000 acres there, yet less than half of this area was in cultivation and receiving water; just five years later, however, by the end of 1920, nearly the entire valley was under crop and receiving water (map 9.1). In order to facilitate the transport of increased quantities of baled cotton to Yuma (fig. 9.2), in 1916 fourteen miles of paved road between Somerton and Yuma were completed. This highway, which ran down the center of the valley, complemented the Yuma Valley Railroad, both of which served as catalysts for the expansion of cotton production and settlement in the valley. Two years earlier, a steel highway bridge had been completed across the Colorado River at Yuma, providing better connections between Yuma and the Indian and Bard units of the project. Realizing the importance of a good network of roads to facilitate marketing, in 1919 voters approved a bond issue to further extend the paving of roads in the region. By 1925, there were sixty miles of hard-surface roads in the Yuma Valley, facilitating the transport of products from farm to markets.[11]

But prosperity at Yuma, as on nearly all the projects during this period, was largely initiated by World War I. From 1913 to 1920, crop prices on government irrigation projects soared, and cultivated land within the projects increased by 241 percent, while irrigated land increased by 77 percent.[12] In the

FIG. 9.1. Short staple cotton in bloom in the Yuma Valley, July 1925. (Reproduced from "Yuma Project Annual Project History and O. and M. Report, 1925")

FIG. 9.2. Cotton gin and baled cotton at Somerton, Yuma Valley, 1924. (Reproduced from "Yuma Project Annual Project History and O. and M. Report, 1924")

case of Yuma, a large number of new settlers moved to the project in 1916 and 1917 to take advantage of the unusually high returns from cotton; some apparently were experienced irrigators from other projects who were seeking a milder winter climate at Yuma.[13] The increased demand for land in the second half of the decade is reflected in the 724 applications received for the opening of 16 reclamation farm units in the Yuma Valley in December 1918. Land values on the project doubled between 1915 and 1918, and in 1919 the average price of improved farmland at Yuma ranged from $350 to $400 per acre, but some parcels were being sold for as much as $500 per acre.[14]

Much of the raw land held for speculation in the Yuma Valley in excess of the allowable 160 acres, and parcels owned by nonresidents, was sold to settlers who brought the land into cultivation as quickly as possible in order to benefit from the prevailing high prices.[15] Since the Reclamation Act prohibited the sale of water rights to nonresidents and to landholdings greater than 160 acres, this was an opportune time to unload these undeveloped tracts at a handsome profit, even though neither provision was ever strictly enforced. In fact, 1918 was the most active year in the sixteen-year history of the Reclamation Service in terms of the transfer of private lands on all the projects.[16] Although a drop in prices paid for cotton and other farm products was experienced in the immediate post–World War I years, cotton prices rebounded in the mid-1920s; as a result, there was little movement among farmers on the Yuma Project toward crop diversification beyond cotton and alfalfa.[17] These two crops continued to dominate farming in the region well into the 1930s.[18]

After the war a readjustment period set in, as average crop values on government projects fell by nearly half between 1919 and 1922.[19] As crop prices plummeted, so did the price of land, and much of the land that was sold at exceedingly high prices on the Yuma Project went back to the former owners or to parties holding the mortgages. The general agricultural conditions on the project in 1922 were very discouraging. Cotton and alfalfa continued to be the principal crops grown, both occupying approximately equal acreage, with their combined total representing 82 percent of the cultivated land at Yuma. Even though cotton prices rebounded somewhat in 1924, land values on the project during the period of agricultural decline dropped precipitously to between $150 and $300 per acre, and they remained at that level for well over a decade. The number of land transfers at Yuma also saw a significant decline from previous years.[20]

But not all those who came to Yuma purchased land; farm tenancy increased in tandem with the expansion of acreage planted to cotton. Of the 1,225 farms counted on the Yuma Project in 1920, 519, or 42 percent, were operated by tenants.[21] Although there had always been a large percentage of tenants on the various Reclamation projects, the figure for Yuma was much higher than the average figure of 24 percent reported for all projects in 1920. The projects with a mild climate where specialized crops were being grown, such as at Yuma, tended to experience higher rates of tenancy.[22]

In the case of Yuma, the high incidence of tenant farming reflected the inability of Yuma Valley landowners to bring their tracts under cultivation fast enough to take advantage of the war-inspired boom, but it was also related to the large number of leased Indian allotments.

The institution of tenancy, however, ran counter to the principal objectives of both the Reclamation Act and the General Allotment Act, but it took on different meanings on Reclamation projects depending on whether it applied to reclamation homesteads or to Indian allotments. In 1924, it was reported that one of the principal needs of the Yuma Project was to increase landownership by those who farmed.[23] This remark was no doubt directed toward land that had been alienated by non-Indians, because tenant farming was viewed as undesirable where the expansion of family-owned and -operated farms was the goal. On the other hand, tenancy was perceived as indispensable on Indian allotments. Government officers tended to encourage leasing, as non-Indians generally were anxious to make use of Indian allotments, and, at least initially, it was far easier to administer leased property than to educate and supply Indians so they could use their own property.[24] Leasing seemed to be the only known way that all of the irrigable land on the Yuma Project could be brought under full production. The double meaning associated with this one feature of reclamation—tenancy—highlights the duplicity with which the federal government pursued its reclamation program at Yuma.

Meanwhile, the increased amount of water being applied to the land in the Yuma Valley resulted in a gradual rise in the groundwater plane across nearly the entire valley.[25] After the drainage debacle in the Bard district, it was evident that the Reclamation Service needed to install drainage works before the water table rose so close to the surface as to render farms unfit for cultivation. Construction work on the main drain in the Yuma Valley was therefore begun as soon as the rising water table was detected. Also, having learned from Bard, much of the land near the levee in the Yuma Valley that potentially could be affected by seepage was never opened to settlement.

One factor that enhanced the effectiveness of drains, both in the Yuma Valley as well as on the California side, was the steady decline in water use among irrigators. When irrigation began to be monitored on the project, an average of just over six acre-feet of water was being applied annually per acre, whereas by 1920 slightly less than three acre-feet of water was used per acre cropped. Water use generally remained at that level through the following decade (table 9.1). Hence, the per-acre usage of water declined

TABLE 9.1
Acreage in cultivation and average water use per acre
on the Yuma Project, 1908–1930

YEAR	AREA CULTIVATED (thousands of acres)	WATER USE (acre-feet per acre)
1908	3.0	6.50
1909	6.5	4.80
1910	10.0	3.10
1911	10.0	5.50
1912	13.5	4.60
1913	19.5	4.40
1914	25.0	3.90
1915	27.5	3.35
1916	30.0	3.25
1917	37.0	3.65
1918	45.5	3.25
1919	54.0	2.80
1920	55.5	2.90
1921	52.5	2.65
1922	54.0	2.55
1923	53.0	2.80
1924	53.0	3.45
1925	56.0	3.05
1926	55.0	2.60
1927	52.5	2.90
1928	54.0	3.05
1929	54.5	2.75
1930	54.5	2.80

SOURCE: "Yuma Project, Arizona-California, Annual Project History, Calendar Year 1935."

as the acreage under crop increased. Originally, the tendency among irrigators was to use too much water partly because of inexperience, partly because of the variable nature of the soils being irrigated, and also because irrigators initially feared that something might interfere with their water deliveries.[26]

The decline in water use over time was due not only to increased confidence in the Reclamation Service to provide timely deliveries and to improved knowledge of irrigation methods; it was also related to the growing acreage planted to cotton and alfalfa seed, crops that required much less water than alfalfa hay.[27] In fact, after a gradual rise in the water table over much of the Yuma Valley since the time intensive irrigation began, it was found in 1919 that the groundwater plane had actually dropped across 60 percent of the area.[28] The lowering of the water table was attributed to the extensive area planted to cotton, to more efficient irrigation methods adopted by irrigators, and to the fact that the annual flood on the Colorado River for the previous couple of years had been below normal.

In 1920 the Yuma Project had a total farm population of 5,100, but in spite of that number farmers' cooperative organizations at Yuma, as on other federal irrigation projects, were habitually slow to develop. This condition prevailed largely because the government retained control of the operation and maintenance of the project works after completing construction, thus preempting the best foundation available for cooperation to develop among the water users. Prior to federal intervention, farmers would band together into mutual irrigation associations not only to construct canals and laterals but also to operate and maintain them. In former times, cooperation was forced on irrigators because it was foolish to try to farm without interacting with one's neighbors. It was this type of interaction that generally led to cooperation in other endeavors vital to the welfare of a community.

Under government patronage, however, this indispensable piece was missing from the reclamation puzzle. In 1914 there were two or three cooperative organizations for producing and marketing cotton and alfalfa seed on the Yuma Project, but the cooperative tendencies were less than satisfactory. Ten years later the Reclamation Service reported that the farmers were exhibiting greater interest in supporting their marketing organizations.[29] Normally, on most of the projects it took ten to fifteen years to build successful cooperative ventures out of which social development and a sense of community spirit evolved. Not coincidentally, it generally took about the same amount of time for most of the projects to develop to the point where they could truly claim successful agriculture.[30]

For two decades following passage of the Reclamation Act, land speculation also plagued the various Reclamation projects. The 160-acre limitation was initially assumed to be an adequate way of keeping speculators in check, but the difficulty and expense of bringing such large parcels under irrigation encouraged farmers simply to hold on to their nonirrigated lands until prices reached inflated values. This situation delayed orderly settlement on most of the projects. As of 1913, the average price of unimproved land had increased by 759 percent on the original twenty-one projects.[31] Land speculation on the Yuma Project during the construction phase, however, was less pronounced, with the average values for unimproved land increasing by less than half as much.

On many projects settlers made their land entries after it became known that a Reclamation project was in the works. Under these circumstances, premature entry and settlement were made largely for speculative purposes. Homesteaders were frequently able to demand high prices for the relinquishment of their homestead right to later bona fide settlers, even though their land remained dry. Consequently, the next wave of settlers, the serious farmers, had hefty financial burdens to bear. Not only would they have to clear and level the land and prepare the soil, while building their houses and barns, and to pay for the construction of the irrigation works, but they would also have to shoulder the burden of paying high interest rates on inflated prices that speculators had charged for the land. Ironically, it was the water furnished by the government that gave value to the land, but because of having to struggle under debt with high interest charges, the prevailing assumption was that the paternalistic government could be put off. This left the government with no alternative but to make concessions in the form of the extension and the graduation of payments.

Those who first made entries in the Yuma Valley, however, did so long before the Reclamation Service surveys were initiated. Here the first wave of settlers were serious pioneers who earnestly attempted to subdue the land through irrigation but were unaware of the difficulties associated with mastering such an immense stream as the Colorado. They did not come to Yuma for speculative purposes, but when land prices began to rise as a result of Reclamation's intervention, they, just as their counterparts elsewhere, took advantage of the escalating land values; landowners who held acreage in excess of the permissible 160 acres, as well as those who held more than they could actually farm, were in a position to make windfall profits during the war years. But the ability of second-wave farmers to carry their repayment obligations to the government was lessened by this type of speculation because land boom prices placed an unbearable hardship on serious irrigators, causing them continually to seek relief from construction charges.

In the case of Bard, many of those who won their farm units in the lottery had no intention of farming and, as we have seen, relinquished their holdings rather quickly. Nevertheless, wholesale profiteering was somewhat controlled here because Bard was not opened for settlement until all irrigation works were complete, and the size of individual farm units did not exceed 40 acres. In fact, in an attempt to regulate speculation, in June 1910, Congress approved an act stipulating that no entrymen would be permitted to file on lands reserved for irrigation purposes until the size of farm unit, the construction charges, and the date of delivery of water had been fixed and publicly announced, just as had been done a few months earlier at Bard. And on the Indian unit of the Reservation Division, the Quechans, of course, were prohibited from selling their allotments, so speculation was not an issue on this unit of the Yuma Project.

The Failure of Allotment

The plight of the Quechans took a different turn from that of the non-Indians who settled project lands. Supporters of the agrarian policy outlined in the General Allotment Act had hoped the measure would accomplish several objectives after its approval: break up Indian communal farming practices and dissolve the tribe as a social unit, secure Indians' individual title to the land, encourage individual initiative, further the progress of Indian farmers while reducing the cost of Indian administration, open unallotted lands to white settlers, and ultimately promote civilization and assimilation.[32] Although allotment failed to achieve most of these goals, it did result in the rapid transfer of surplus Indian land to non-Indian settlers, and, through allotment in severalty, it diminished the influence of the tribe as a social unit. By 1934, only 52 million acres remained in tribal hands out of an original 130 million acres prior to approval of the allotment policy. Of this acreage, 66 percent was still communally owned by the tribes, while 34 percent was allotted in severalty.[33] As on other allotted reservations, the assignment of allotments to the Quechans was the first step in what the government had hoped would be a long-range program of agricultural development as the primary economic base for the Quechan community. Unfortunately, that did not happen at Yuma, nor did it occur in the majority of cases. In the case of the Quechans, inadequate government support and the presence of land-hungry non-Indians anxious to lease Indian allotments, as well as cultural resistance to irrigated farming, conspired against the Indian allottees becoming successful agriculturalists.

In 1925, local Reclamation authorities ominously reported that the Indian lands did not show the care of previous years.[34] At that time, approximately 6,750 acres out of 8,100 acres of Indian allotments were irrigated. Many of the improvement leases on the reservation had now expired, and the Quechans were being encouraged to farm their own allotments. But the amount of irrigated land on the reservation continued to dwindle. By 1932, only 2,991 acres of Indian allotments were being cultivated, 1,106 by lessees and the remainder by the Quechans.[35] In 1940, the area under cultivation amounted to only 2,246 acres, with the Indians farming 1,243 acres and the lessees 1,003 acres.[36] Commercial crops, mostly alfalfa and cotton, accounted for nearly all the leased acreage, while about half of the land worked by the Indians was devoted either to subsistence crops such as maize, beans, and garden crops or to pasture.

Part of the decrease in acreage under cultivation was caused by the Great Depression years, resulting in a significant decline in the demand for land by lessees; part also was due to the damage that had been inflicted on Indian allotments by seepage and alkali. In 1933, 1,425 acres of the Indian unit were classified as nonirrigable because of alkali deposits resulting from a high water table and nonuse. Non-Indian farmers on the reservation, in their haste to make a profit off the Indian allotments, repeatedly overirrigated, causing a rapid rise in the water-table plane. Since lessees had no vested interest in the land itself because their five-year leases precluded them from making any long-term improvements, they sought to make as large a profit as possible before their leases expired. Then when farming was abandoned, alkali concentrations began to accumulate. And to further compound the problem, in 1939, seepage from the newly completed All-American Canal, which diverted water from the Colorado River at Imperial Dam just above Laguna Dam and channeled it along the edge of the mesa north of the floodplain on its way to the Imperial and Coachella valleys, intensified the groundwater problems and salinization on the reservation. A report on drainage conditions on the Indian unit in the 1940s stated that the most important work needed to bring the irrigable lands of the reservation into proper use was drainage.[37] By then approximately 4,000 acres, or half the Indian allotments, had stood idle for nearly a decade.

Besides land degradation, a major obstacle to the ongoing development of the Indian allotments involved the tenure status of Quechan lands. We have seen that non-Indian homesteaders were restricted from mortgaging their holdings to raise much-needed cash to make necessary improvements, but this constraint prevailed only while proving up, initially five years, later reduced to three years. After receiving title, homesteaders were free to use their land as collateral for investment capital. The Quechan allotments, on the other hand, were given individual trust status for twenty-five years, and, with ratification of the Indian Reorganization Act in 1934, the trust period on their land was extended indefinitely. The latter action served to guarantee that the Quechans would lose no more of their reservation, but the trust patents forever constrained the Quechans' ability to use their land as collateral. In other words, Quechan titles were inalienable and therefore could not readily be leveraged for capital in a capital-intensive farming environment. Without the ability to take out mortgages, the Quechans could

TABLE 9.2
Number of Quechan owners per ten-acre allotment, June 2000

NUMBER OF OWNERS	NUMBER OF TEN-ACRE TRACTS	NUMBER OF OWNERS	NUMBER OF TEN-ACRE TRACTS
1	10	30	6
2	149	32	4
3	64	33	2
4	62	35	2
5	57	36	7
6	36	37	2
7	37	38	3
8	52	39	2
9	20	40	1
10	18	41	6
11	28	42	9
12	32	44	4
13	14	46	4
14	14	47	4
15	10	48	1
16	12	50	1
17	11	53	2
18	11	54	2
19	9	55	4
20	13	62	1
21	18	64	1
22	7	67	1
23	11	72	1
24	13	76	1
25	9	80	1
26	6	85	1
27	5	104	1
28	9	109	1
29	5	127	1

SOURCE: Department of the Interior, Bureau of Indian Affairs, Land Titles and Records Office, Albuquerque.

not generate the necessary funds to farm their holdings even if they had an interest in doing so.

Additionally, equal-share inheritance to the descendants of allottees resulted over time in complex ownership patterns. Originally, the 10-acre individual allotments gave Quechan families on average about 30 acres, although 38 percent of the family allotments were only 10 acres in size. Descendants of those families who received a sufficient parcel size for farming at the time found that heirship subsequently interfered seriously with the operation of the land in suitable farm units, forcing the Quechans to become more, rather than less, dependent on government support and assistance. In the year 2000, nearly ninety years after the allotments were parceled out, only 10 of the original 819 allotments had single owners. Seven or fewer individuals share approximately half of the allotments, with two owners per tract being the most common (table 9.2). However, 30 owners or more are found on 76 parcels, and one 10-acre allotment is shared by as many as 127 individuals.[38] And many Quechans own shares in several separate plots. Under these circumstances, none of the successors could make a profit from cooperatively farming their own holdings.

Leasing and other land-use decisions were also complicated by this arrangement because they required unanimous agreement by all landholders, a task that became more difficult as heirship increased.[39] As a result, many undivided heirship lands sat idle and returned to their unimproved state because landowners were unable to reach a consensus regarding how the land should be farmed. The allotments, which in time were held by so many owners in common, rendered the Indians helpless to make effective use of their property. An unanticipated turn of events had caused the Quechans to become less self-sufficient than they had been before allotment. Under the best of circumstances, the General Allotment Act had forced the Indians to become petty landlords who were largely dependent on unearned income from the only commodity they owned—their land.

"A Fundamental Error"

In the year following passage of the Reclamation Act, chief engineer Newell wrote, "The object of the reclamation law is primarily to put the public domain into the hands of small land owners—men who live upon the land, support themselves, [and] make prosperous homes."[40] Newell's principal responsibility, therefore, was to create additional opportunities for homesteading in the arid West—to convert the desert tracts that could not easily be irrigated by group effort or private irrigation ventures into self-supporting agricultural communities.

In spite of this noble goal, little besides the available water supply and the engineering features were considered in the authorization of the initial set of projects, whereas their agricultural and economic feasibility was a matter receiving little attention in the early days of federal reclamation. Instead, those who came to take up reclamation homesteads were for the most part unqualified and unprepared to carry out the difficult and costly work of irrigated farming, and many had no serious purpose other than land speculation. From an engineering standpoint, the irrigation works built by the Reclamation Service were both impressive and functional, but by the 1920s it was widely recognized that "a fundamental error was made in believing that the construction of irrigation works would of itself create irrigated agriculture."[41]

In the 1920s, federal reclamation project acreage amounted to 1.25 million acres, just a fraction of the 7.5 million acres of farmland already irrigated in the West in the early 1900s.[42] Attention had been centered almost entirely on engineering features, as Reclamation Service engineers were both eager and ego driven to complete the necessary irrigation works regardless of the natural deterrents and the costs involved, and were therefore heedless to the human side of reclamation. Little or no consideration was given to the potential obstacles that settlers might encounter in bringing the land under cultivation, equipping their farms, and meeting construction charges.

This oversight no doubt was at least partly related to the knowledge that the cooperative efforts of settlers on arid lands in the past had successfully brought desert tracts under irrigation. All the government had to do, it was thought, was provide the "artificial streams" that homesteaders could tap into, and the rest would fall into place. Twenty years later, Newell, now working as a consulting engineer for the Reclamation Service, admitted this failing. In 1922 he wrote that "the human problems, those of the selection of settlers, ways of cultivation of the soil, the maintenance and increase of its fertility, the proper selection and rotation of crops, marketing, cooperative buying and selling, and particularly the social relations, should be the subject of the same kind of scientific research and of practical application of results as has been found essential in handling the engineering or material side of the problem."[43]

Introspection

Because of the less than satisfactory results of Reclamation policy and the failure of the policies of the Indian Service to advance the socioeconomic status of Native Americans, the mid-1920s became a period of federal introspection and soul-searching. Accordingly, in 1923, Interior Secretary Hubert Work appointed a "Committee of Special Advisors on Reclamation" to make an intensive study of national reclamation policy and to develop a series of recommendations that would bring a higher level of success to national irrigation projects.[44] Three years later, in the spring of 1926, Secretary Work enlisted the Institute for Governmental Research to undertake an independent and impartial survey of the general economic conditions of the Indians. The resulting document, which came to be known as the Meriam Report, chronicled the plight of Native Americans in the 1920s and denounced the policies responsible for the prevailing situation.[45]

The Committee of Special Advisors on Reclamation consisted of six members who were experts on irrigation matters or possessed considerable knowledge of conditions in the West, one of whom was Elwood Mead. Mead, it will be recalled, chaired the local board of review at Yuma a decade earlier, and therefore had intimate knowledge of the problems associated with the Yuma Project. By now, Mead had become the foremost professional name in irrigation in the country. The committee concluded that federal reclamation policy had failed to accomplish the human and economic purposes for which it was created.[46] Settlers without much capital who were inexperienced in the ways of irrigated farming, instead of being encouraged to work together, were left to struggle without sufficient aid or direction to complete what the government had begun.[47] Included among the litany of shortcomings in the committee's final report were the following: settlers should have been screened according to their farming background and financial solvency; the heterogeneous body of water users should have been given more and better information by competent official advisers; farmers with little capital should have been provided with proper credit facilities; farm units should have been surveyed with an eye toward their productivity, and hence their ability to pay construction charges; speculation should have been prevented; and assistance should have been provided in the establishment of cooperative organizations for buying and selling.

It was also implied that a dependence on federal paternalism had become a common feature on nearly all the projects, and that the water users had come to look upon themselves as "wards of the government," a perception that reclamation policy unfortunately had theretofore tended to encourage.[48] Some of these problems might have been alleviated if a more gradual and orderly program of development had been adopted allowing for the steady accrual of knowledge of the federal irrigation venture, instead of pursuing the construction of nearly two dozen projects practically simultaneously. In fact, many of the difficulties that the Reclamation Service faced could be traced to the initial haste and the vastness of its program. Federal irrigation, it seems, was an experiment that had sprouted before it had fully taken root.

The Meriam Report examined the wide-ranging problems of Indian administration, but paid particular attention to the relationship among Indian economic resources, land tenure, and socioeconomic well-being. The report was unrelenting in its attack on the allotment program and was the first major statement advocating a new policy focusing on the consolidation of tribal efforts as a way to improve the socioeconomic status of Indians.[49] The government had hoped that through the partition of land in severalty, which was analogous to the granting of a homestead, tribal associations would be dissolved and individual status would be substituted in its place. Although this did not happen in most cases, the allotment system generally did weaken the cohesion of the one cooperating unit to which the Indians were culturally attuned—the tribe. By undermining the Indian way of life, allotment subverted the only sound foundation upon which a transformed Indian society could have been built. Additionally, even though individual allotments were normally grouped in family units, as in the case of the Quechans, they nevertheless were made to the individual and not to the family. This contrasted with the traditional homestead that embodied the family farm tradition. Consequently, allotment tended to shatter family interests as well, and to work as a divisive factor in the lives of all allotted Indians.

The allotment policy therefore did just the opposite of what was intended. The Indians generally did not know the value of their land, they did not know how to use it, and the government as a rule did not supply persons competent to teach them its use. Since many Indians were not ready to make effective use of their individual allotments, the solution was to permit them to lease their lands to non-Indian farmers. But the paltry lease money gave the Indians unearned income that invited a life of idleness, generally at a dangerously low standard of living, instead of forcing them to face the necessity of working. Compounding this situation were the proceeds received from the sale of surplus reservation land after the allotments were made. This money was placed in a tribal fund, from which periodic payments would be made to tribal members. These annuity payments only added to the Indians' unearned income and postponed the day when it would be necessary for them to seek secure employment to support themselves. In the case of the Quechans, the average annual per capita income from individual and tribal sources among members of the tribe in the mid-1920s was reported to be a mere $97.[50]

Despite the failure of allotment, the Meriam Report conceded that agriculture remained the chief economic possibility for the majority of Indians since their principal economic resource was their land. However, it recommended that the effort be directed away from competitive market agriculture to the development of subsistence-oriented family farms. In places where the allotment policy became entangled in Indian irrigation, as at Yuma, the situation was more complex because the Indians needed to be trained in sophisticated irrigation methods so that they could eventually adopt commercial farming, a system of agriculture that implied an ability to buy and sell and transact business, activities for which the Indians had little interest, aptitude, or experience. Given the composition of the Meriam Commission, with its lack of irrigation specialists, Interior Secretary Work, in 1928, asked the secretary of agriculture to convene a second committee to look strictly at the status of irrigation on Indian reservations. This committee comprised Porter J. Preston, an engineer from the Bureau of Reclamation (formerly the Reclamation Service), and Charles A. Engle, a supervising engineer from the Bureau of Indian Affairs. Preston and Engle concluded, "It cannot be said that there has been any serious and sustained effort on our part to make farmers of the Indians." But the fundamental defect of the government's agrarian program aimed at Indians, and hence allotment, was that it was based on the assumption that it was possible to make every Indian a farmer, an objective no more possible or desirable among Indians than it was among non-Indians.[51]

A quarter century prior to the publication of the Meriam Report, the commissioner of Indian affairs noted in his annual report that "the Indian has been allotted and then allowed to turn over his land to the whites and go on his aimless way. This pernicious practice is the direct growth of vicious legislation." Between 1887, when the General Allotment Act was approved, and 1934, when it was superseded by the Indian Reorganization Act, approximately 80 million acres had passed out of Indian ownership, leaving approximately one hundred thousand Indians landless.[52] Leasing to non-Indians, as we have seen, was another means of securing the use of Indian land, if not title to it. This was especially true on so-called Indian irrigation projects, where, in the late 1920s, most farming was being done by non-Indians.[53] Whether the loss of use of Indian land through sale or lease was due to "vicious legislation" or was the product of congressional actions that underestimated the enormous cultural chasm that existed between the country's Euro-American majority and its Native American population is open to debate. What is obvious, however, is that policy makers "refused to consider seriously the land and peoples they were so confident they could remake."[54]

Yuma Crossroads

Many, if not most, of the observations found in the reports commissioned by Interior Secretary Work in the 1920s would have been evident after an intensive study of developments associated with the impacts of federal intervention at Yuma. Yuma, in this sense, served as an important crossroads where legislation involving both federal irrigation and Indian allotment intersected. Yuma was unique among Reclamation projects, for this tiny corner of the Southwest encompassed nearly all of the problems and pitfalls connected with the federal government's attempts to convert the nation's final frontier into irrigated oases. Since the Fort Yuma Indian Reservation was entangled in the development of the first Reclamation project intended for white settlement on the lower Colorado River, this detailed examination of the Yuma Project highlights the unforeseen difficulties associated with federal involvement in bringing western lands under irrigation and settlement by non-Indians and Indians alike. It also underscores the unanticipated engineering problems encountered by the Reclamation Service in its attempt to control a large and tenacious stream in the arid West.

Despite all the mistakes, federal involvement in irrigation ultimately did generate additional opportunities for homesteading in the West, partly because of the newly created opportunities to exploit Indian natural resources. Paradoxically, although the Yuma Project might be exceptional because of its multifaceted nature, at the same time it demonstrates the capacity of local case studies to enhance our understanding of the settlement processes that characterize the broader region in which they are situated. Yuma, therefore, is paradigmatic of the range of difficulties associated with settling the nation's final frontier.

Although the collection of federal agrarian policies designed to tame the final frontier were seemingly well intentioned in the abstract, they were misguided in their execution. They reflected the lack of congressional understanding involving the realities of western settlement and the cultural beliefs and traditional livelihood patterns of Native Americans. Most important, most federal policies were anathema to community cohesion, the cornerstone of successful irrigated farming in the arid West. The results of Reclamation policy in its early years were disastrous largely because the government failed to recognize the importance of economic and educational assistance and the development of a cooperative spirit on project lands, whereas the allotment policy, partly for the same reasons, ultimately led to a paralysis of productive Indian land. Both policies, because of their paternalistic nature, tended to foster dependence among their respective subjects.

Related to the bureaucratic blunders that made the development of successful irrigated agriculture so problematic were the unanticipated or simply overlooked environmental impacts that plagued the Yuma Project during

FIG. 9.3. Lettuce production on a Quechan allotment in the Indian unit of the Yuma Project. (Photo by the author)

the early years. The struggle to tame the Colorado River turned out to be a much more difficult task than expected. In fact, the lower Colorado was never brought fully under control until the 1930s with the completion of Hoover Dam. After the river was leveed, the rise in the water table during flood stage had near catastrophic consequences. Although government soil scientists predicted this circumstance even before the approval of the Yuma Project, Reclamation Service engineers, who focused solely on the works they were creating, paid little heed to this potential environmental repercussion. And although the importance of drainage was also apparent before the approval of the Yuma Project, it was largely overlooked in the early years of settlement and development. Finally, no attention whatsoever was given to the varying quality of soils found at Yuma, a circumstance that would eventually cause considerable dissatisfaction among settlers on the project.

Considering the fact that the Reclamation and General Allotment acts met head-on at Yuma, and given the environmental problems that characterized the region, one can't help but marvel at the intensity and diversity of the farming landscape etched into the floodplain around Yuma today (figs. 9.3, 9.4, and 9.5). However, although irrigated farming on the Yuma Project is highly sophisticated and extraordinarily productive, it is now dominated by corporate enterprise. As Donald Pisani has noted, "Far from being a revolu-

FIG. 9.4. Date orchard in the Bard unit of the Yuma Project. (Photo by the author)

FIG. 9.5. Broccoli production in the Yuma Valley unit of the Yuma Project. (Photo by the author)

tionary tool for social change, irrigation became and remained the servant of 'agribusiness.'"[55] Both the Quechan allottees and the non-Indian farm unit owners on the Yuma Project, with their irrigable, highly fertile floodplain farming tracts distinguished by a virtually year-round growing season, have experienced the same fate as other small American landowners caught up in the tidal wave of corporate farming.

Hence, today, nearly all the Quechan allotments are indirectly leased to large produce corporations, such as Dole. The local Bureau of Indian Affairs office arranges the leases and handles the complex bookkeeping involved in distributing payments among the multiple owners of each plot. Although many Quechans still live on their allotments (53 percent of the tribe's members live on or near the Fort Yuma Reservation), they do not farm, save for some kitchen gardens near their homes.[56] Inadequate government support, the prevalence of non-Indians anxious to lease, cultural resistance, and the increasingly corporate nature of American agriculture have all conspired against the Quechan allottees becoming farmers. The reservation Indians do not even work as farm laborers anymore, having been supplanted by Mexican day laborers. Like many other Indian tribes, the Quechans have turned to casino gambling as a major source of employment and income. In 1996, a small casino was opened on the Fort Yuma Indian Reservation. Today, two highly profitable casinos connected by a walkway (one in California, the other in Arizona) generate an estimated $45 million in net revenue per year for approximately 3,300 Quechans.[57] Although the gambling operations employ about 28 percent of the tribe's people, the Quechans still struggle with a 67 percent unemployment rate.[58]

The majority of Bard unit owners, as well as those in the Yuma Valley, also lease their land to large produce corporations, while many of the owners hire out their services as paid growers. Most owners simply do not have the resources and the fortitude to face fluctuating market prices or uncertain weather conditions.[59] Highly capitalized national corporations, on the other hand, are able to absorb temporary losses when necessary.

As one scans the verdant Yuma panorama from Indian Hill today, it is difficult to imagine the symbiotic relationship that once existed between the Quechans and the Colorado River as the Indians carried out their flood recession agriculture and other subsistence activities in a land they once dominated. David Rich Lewis has noted that subsistence was a "core" structure of all Indian cultures, and that most Native Americans tended to practice productive subsistence security rather than the productive maximization of their environments.[60] In doing so, each group utilized a diverse range of resources rather than concentrating on a single commodity or mode of production. The Quechans were no different. But access to diverse resources became more limited as the Quechans were herded onto their reserve, awarded individual allotments, and stripped of surplus reservation land.

The Quechans, and other Native American groups, generally resisted elements of white society that required radical change. *Radical* does not begin to describe the agrarian policy imposed on the Quechans requiring the adoption of sophisticated irrigation techniques and market-oriented agriculture. The government's failure to take into consideration the traditions and successful subsistence adaptations of the Quechans before instituting wholesale change, by attempting to force on them irrigated agriculture, for example, disregarded cultural, economic, and environmental realities that ultimately limited their participation in the agrarian program designed to help them. Extrapolating similar practices to other Indian groups, it is not surprising to find that Indian farming actually declined following allotment because few groups were willing to abandon completely core elements of their culture, including diversified subsistence practices.[61]

It is also difficult to conjure from Indian Hill the myriad complications connected with the establishment of industrial irrigation at Yuma so that successful agriculture could eventually take root there. Nor are the environmental repercussions connected with early irrigation developments and the distresses suffered by settlers apparent today. These features connected with Yuma's past have largely been vanquished and expunged from the present-day scene. Hence, one might argue that in the final analysis the government was successful in achieving its goal of converting Yuma Project lands into an intensively cultivated, high-value, irrigated farming district.

On the other hand, federal reclamation has taken on a much different image than that originally intended by indirectly providing large produce corporations with subsidized water in order to grow high-value crops. This obviously is not what the framers of the Reclamation Act had in mind, for the small family-owned and -operated farm that federal reclamation was designed to promote is only a memory today. Meanwhile, the legacy of allotment survives, for the Quechans continue to struggle under the burden of a failed federal policy that moved them, and many other Native Americans, "from cultural self-sufficiency to dependency and enforced marginality," although the expansion of casino gambling on the Fort Yuma Reservation provides more promise for the tribe's economic development in the future.[62] Even though today the intensity, the value, and the variety of agricultural production at Yuma are impressive, and most likely would not have been possible without federal involvement, when viewed from the standpoint of the original federal objective, federal reclamation and allotment at Yuma cannot be deemed an unqualified success.

Epilogue

It is along the Colorado River that "the most celebrated showpieces of American water engineering have appeared."[1] In the process of furnishing water to nearly 30 million people and 3.7 million acres of farmland in seven western states and northern Mexico, the Colorado has become one of the most dammed, diverted, and (over)regulated rivers in the land; in fact, it is "a river no more," according to Philip Fradkin, who documented the Colorado's decline from its headwaters to the Gulf of California.[2]

We have seen that the river's transformation began more than a century ago near Yuma with the construction of Laguna Dam. This early irrigation project would lead to subsequent developments on the Colorado that would result in unforeseen troubles decades later. Excluding Laguna, eight dams now straddle the Colorado; some serve to regulate the river's flow, while others function to divert its flow. Today, Laguna Dam has been reduced to a regulating structure for the small reservoir found behind it, which is now used for recreation. Laguna's utility as a diversion structure for the Yuma Project has been supplanted by Imperial Dam, completed in 1939 a few miles upriver, which diverts Colorado River water into the All-American Canal, serving the Imperial and Coachella valleys, as well as the Yuma Project. The more noteworthy issues and concerns involving the subsequent storage and diversion projects on the Colorado are the subject of this final chapter.

The All-American Canal

The great dam-building era and ensuing activities associated with industrial irrigation on the Colorado grew out of the need to regulate the cycles of flood and drought that characterized the annual flows of the river. Uncontrolled, the Colorado was of limited value, particularly in the rapidly growing Imperial Valley, where sediment deposits and irregular flows compromised water deliveries.[3] Laguna Dam, then serving as a diversion structure for the

Yuma Project, was not designed to regulate the lower Colorado River's flow. Imperial Valley interests, who were not anxious to experience another disaster similar to the one they had endured from 1905 to 1907 during the Colorado River breach, began to campaign for the construction of a storage dam to control the river, and a canal that would carry water from the Colorado River to the valley without crossing the border into Mexico. In 1919, the Imperial Irrigation District (successor to the California Development Company that purchased the defunct corporation's irrigation works) forged a partnership with the long-resisted Reclamation Service to determine the feasibility of constructing an "All-American Canal" to carry water from the Colorado River to the Imperial Valley. Consequently, a comprehensive survey of the lower Colorado River basin was undertaken, which culminated in the so-called Fall-Davis Report, submitted to Congress in 1922.

Arthur P. Davis, who, as noted earlier, was fixated on the conquest of the Colorado River, now served as director of the Reclamation Service. His agency was responsible for conducting the survey of the lower Colorado River, the findings of which were turned over to Interior Secretary Albert Fall. The resulting "Fall-Davis Report" recommended a comprehensive development scheme to regulate the river, generate power, and irrigate the lower Colorado Valley.[4] In order to accomplish this, Davis advocated approval of the Boulder Canyon Project involving the construction of a large storage dam, later to become known as Hoover Dam, in the vicinity of Boulder Canyon, and the completion of an "All-American Canal" leading from the Colorado River to the Imperial Valley, bypassing Mexican territory. Twenty years earlier, Davis had recognized the need for such a dam as part of a comprehensive plan to develop the river. The dam would be paid for by leasing the right to generate power, while the reservoir behind it would lessen siltation and prevent major flooding downstream, and ensure the availability of water to growers.

The All-American Canal was estimated to cost thirty million dollars, but its repayment by Imperial Valley irrigators would be spread over many years interest free, in accordance with the 1902 Reclamation Act and its subsequent amendments.[5] And most of the money would come from power revenues collected by the Imperial Irrigation District from generating facilities located along drops in the canal. Consideration was given to using the existing Laguna Dam to divert water into the new canal, but it was decided to build a new diversion structure, Imperial Dam, further upriver. The heading of the proposed Imperial Dam was twenty-one feet higher than that found at Laguna Dam, thereby reducing the time and cost of excavating the canal across Imperial Dunes, the sand hills west of Yuma. Congressional approval of the Boulder Canyon Project Act ushered in "an era of massive federal water projects that transformed the Colorado into its highly controlled state." It also authorized California, Arizona, and Nevada to enter into a compact to apportion the Colorado's lower basin water, and required California to limit its right to 4.4 million acre-feet of Colorado River water, plus one-half of any available surplus in a given year.[6]

The Colorado River Compact

Although the engineering and economic feasibility of the above projects had now been established, interstate water disputes and other factors delayed authorization of the Boulder Canyon Project until 1928. The unexpected friendship between Imperial Valley interests and the federal government that had developed in preceding years caused concern among upper Colorado River basin states, "which saw the water originating in their mountains disappearing into the insatiable maw of California."[7] The Colorado Development Company alone had the legal right to a flow that, at the time, amounted to a substantial portion of the average annual flow of the river at Yuma. Fearing permanent control of the river because of prior appropriation by California, Colorado and other slower-growing upper basin states sought to protect their future water rights by initiating, in 1919, negotiations for an agreement that would assign a portion of the Colorado's flow to each of the seven basin states.

The political reality of the situation was that Imperial interests knew that the Boulder Canyon Project, including the All-American Canal, would never be approved without the support of upper basin states, whose congressmen were on key committees.[8] The federal government entered the negotiations in 1921, and on November 24, 1922, the Colorado River Compact was signed at Santa Fe, New Mexico. The compromise divided the river's drainage into an upper and a lower basin, with the boundary between the two basins located at Lees Ferry, a few miles south of the Utah-Arizona border (map 10.1). At the time of the agreement, the average annual flow of the river at Lees Ferry was estimated to be 17 million acre-feet. Fifteen million acre-feet of the river's flow would be divided evenly between the upper and lower basins. The 7.5 million acre-feet allocated to each basin would be further apportioned among each state within its respective basin. California, Nevada, and Arizona fell into the lower basin, while Colorado, Utah,

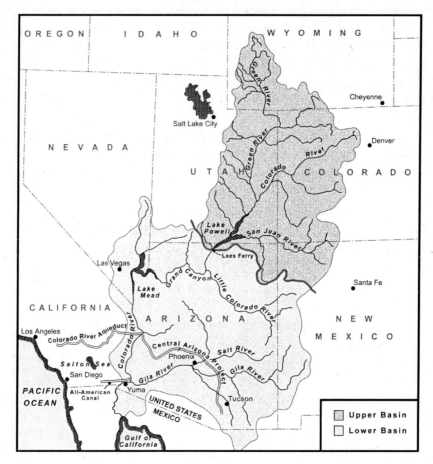

MAP 10.1. The upper and lower basins of the Colorado River watershed.

Wyoming, New Mexico, and the northeast corner of Arizona encompassed the upper basin.

Soon after the Colorado River Compact was approved by the seven basin states, Arizona began to rethink the agreement and refused to ratify the compact. From Arizona's standpoint, the issue of California's appetite for water had not been solved, for although the flow of the Colorado had been divided between the two basins, it had not as yet been divided among the individual states. Although the upper basin was now protected against California's voraciousness, Arizona was not.[9] Arizona eventually ratified the Colorado River Compact in 1944, but the state's share of the river was not finally determined until 1963 after the U.S. Supreme Court's landmark ruling in *Arizona v. California* divided the Colorado's flow among the three lower basin states. California had claimed an annual allocation of 5.4 million acre-feet, based on senior and superior rights, because Arizona had failed to ratify the 1922 Colorado River Compact. The Supreme Court ruling confirmed Arizona's right to 2.8 million acre-feet and limited California to a firm allocation of 4.4 million acre-feet, still well over half of the lower basin's allotment, thanks largely to the prior appropriations made by Imperial Valley interests in 1899.[10] Nevada was allotted the remainder of the lower basin's share, or 0.3 million acre-feet. Meanwhile, the four upper basin states, with little conflict, in 1948 signed the Upper Colorado River Basin Compact, which apportioned their allotments according to the percentage of water available in any given year, instead of by amount, which obviously could vary from year to year. Colorado, the most populous of the four states, received the largest share in the upper basin with 51.75 percent, followed by Utah's 23 percent, Wyoming's 14 percent, and New Mexico's 11.25 percent. Since the northeast corner of Arizona is located within the upper basin's drainage, it received 500,000 acre-feet from that basin's allotment in addition to its 2.8 million acre-feet lower basin allocation.

The Mexico-U.S. Water Treaty

Mexico, lacking a treaty with the United States specifying the amount of Colorado River water it was entitled to, had an understandable interest in the Colorado River Compact negotiations since the Mexicali Valley, the primary agricultural district found in the Mexican portion of the delta, also relied on Colorado River water for its survival, even though the majority of water destined for Mexico was composed of return flow from American irrigators. The United States, however, ignored Mexico's request to participate in the negotiations. Instead, the State Department invoked the precedent of the Harmon Doctrine, named after former Attorney General Judson Harmon, which dealt with a similar matter involving the Rio Grande. The Harmon Doctrine represented "the height of international arrogance," for it asserted the self-serving principle of America's absolute control over water originating upstream from a foreign country.[11] Although the Colorado River Compact made a vague reference to the possibility of a treaty with Mexico at some future date, it left it up to the basin states to furnish Mexico with water. Until a treaty could be negotiated, Mexico would have to continue to be satisfied with its historic entitlement from the Imperial Canal as well as with whatever surplus water might flow downstream in any given year.

Approval of the Boulder Canyon Project in December 1928 further persuaded Mexico to seek a guaranteed treaty for water from the Colorado.
Mexico duly feared that the construction of Hoover and Imperial dams, and
the completion of the All-American Canal, would disrupt the river's regime
downstream. In February 1935, Colorado River water began to back up
against Hoover Dam, creating a massive storage reservoir, Lake Mead. As
Hoover Dam was nearing completion, work began in 1934 on Imperial Dam
and the All-American Canal. The first water was delivered to the Imperial
Valley via the All-American Canal in 1940, but it was not until 1942 that the
canal would flow at full capacity (fig. 10.1).

Meanwhile, to counter the Colorado River developments taking place
north of the border, Mexico encouraged the massive expansion of farming
in the Mexicali Valley, believing that at some point the United States would
have to recognize Mexico's rights to a portion of the river's flow.[12] Once the
All-American Canal was operating at full capacity in 1942, the Imperial Irrigation District no longer relied on the Imperial Canal to carry water from
the Colorado River to the Imperial Valley; the flow of water through the
former canal was, therefore, diminished, and the Mexicali Valley found itself
with insufficient water for irrigation purposes. Consequently, in the spring
of 1943, farmers in the Mexicali Valley were forced to purchase water from
the Imperial Irrigation District at exorbitant prices. Since the United States
now had control over the river's regime, Mexico threatened to take the matter to the World Court, or to submit to international arbitration. Desiring
good relations with its neighbor, especially during the war years, the United
States finally entered into an agreement that guaranteed Mexico a portion of
the Colorado River's flow, or, more correctly, its return flow.[13]

The Mexico-U.S. Water Treaty was signed by the two countries in 1944,
and was ratified by the United States Senate in April 1945. The quantity of
water owed Mexico was set at a minimum of 1.5 million acre-feet per year,
absent extraordinary drought or serious accident, whereas delivery of up to
1.7 million acre-feet is required in surplus years. The treaty also established
the International Boundary and Water Commission "to assist in the management of U.S. and Mexican trans-border resources."[14] The Mexico-U.S. Water
Treaty set the stage for the construction of Morelos Dam on the Colorado
River just below the California border in 1950. Morelos Dam was designed
to capture drainage from the United States and divert it to the Mexicali Valley through the Alamo (formerly Imperial) Canal. The treaty also prompted
the purchase by Mexico of the Imperial Irrigation District's former water-

FIG. 10.1. View of the All-American Canal at Imperial Dunes. (Photo by the author)

works infrastructure for Mexicali Valley farmers. The Mexico–U.S. Water Treaty, therefore, not only guaranteed Mexicali farmers a specific amount of water each year but also freed them from the "capricious control the IID exercised over Mexico's share of water."[15]

It was the impending agreement between the United States and Mexico for the delivery of Colorado River water that encouraged Arizona to ratify the Colorado River Compact in 1944. Arizona feared that if it failed to endorse the compact before a treaty was signed with Mexico, California would insist that Mexico be furnished part of the Colorado River's flow that had been set aside for Arizona. By this time, California already had the capability, if not the legal right, to divert as much as 5.6 million acre-feet from the Colorado River. In addition to Imperial Dam and the All-American Canal, Parker Dam and the Colorado River Aqueduct had by now been completed, which diverted Colorado River water to coastal Southern California. The 242-mile canal, funded by the Metropolitan Water District of Southern California, began carrying water in 1941. Hence, Arizona's motives were far from altruistic in its support of Mexico's claim to Colorado River water. Arizona felt that if a legal limit was not placed on Mexico's share of the river, then sufficient water might not be available to develop its own anticipated water projects. Once Arizona's governor signed the Colorado River

Compact in 1944, the state's senators then began to rally support for ratification of a water treaty with Mexico.[16]

The Gila Project

The Gila Project, the only Arizona water project on the Colorado River that was under development at the time, evolved out of the original Yuma Project, and its approval essentially represented congressional compensation to Arizona for dropping its initial objections to the construction of Parker Dam and the Colorado River Aqueduct. Authorized in 1935, the Gila Project was designed to irrigate approximately 150,000 acres on Yuma Mesa, but construction was delayed during the war years. In 1947, Congress enacted legislation that substantially modified the original project's irrigable land area, reducing it to approximately 40,000 acres, divided between 25,000 acres on Yuma Mesa and 15,000 acres in the North and South Gila valleys (the latter eliminated years earlier from the Yuma Project because of their flood-prone nature and poor drainage). The 25,000 acres on Yuma Mesa would supplement the 3,400 acres already under cultivation in connection with the Yuma Auxiliary Project. A new unit in the Gila Valley, the Wellton-Mohawk Division, named after two small settlements east of Yuma, substituted for the acreage eliminated from the original Gila Project.

The Wellton-Mohawk Division begins about 12 miles east of Yuma and extends on both sides of the Gila River for about 45 miles. It comprises approximately 75,000 irrigable acres. Each division of the project would be supplied with water diverted via the Gila Gravity Main Canal on the Arizona side of Imperial Dam. By 1950 the Gila Project was feeding water to Yuma Mesa and the North and South Gila valleys, and two years later the Wellton-Mohawk Division of the project began to receive Colorado River water. But when the fresh supply of water was delivered to Wellton-Mohawk, it detonated an ecological time bomb that would strain U.S.-Mexican relations for more than a decade.

As we have seen in an earlier chapter, the initial irrigation ventures located near the Colorado-Gila confluence were found along the Gila River. After the Gila River floodplain was opened to settlement, irrigators moved in, raising alfalfa, cotton, and some vegetables. However, calamitous floods that washed out crops and destroyed diversion works, alternating with long periods of drought, caused settlers to begin pumping water from underground sources to irrigate their crops. In 1928, Coolidge Dam, a Bureau of Indian Affairs project located on the Gila River 250 miles upriver from Yuma, was completed. This, coupled with drought conditions, caused the Gila River to carry insufficient water to replenish the wells in the Wellton-Mohawk region. Poor drainage conditions in the valley caused excess irrigation water to seep back into the wells, and by the early 1930s the recycled groundwater had become too saline for most crops.[17] As water and soil became too toxic for successful farming, farms began to be abandoned.

It was the deteriorating situation in the Gila Valley east of Yuma that caused Arizona boosters in 1941 to lobby Congress to incorporate the needs of Wellton-Mohawk irrigators into the Gila Project. The revised project would supply fresh Colorado River water to the region, thereby releasing irrigators from their reliance on contaminated well water. When the first water deliveries from the Colorado began in 1952, it initially resulted in a rapid expansion in cultivated acreage. But as copious supplies of Colorado River water were applied to the land, coupled with the poor drainage conditions, the saline-saturated water table began to rise at an alarming rate, threatening farmers' crops. Drainage wells were constructed to remove the excess groundwater, but the wastewater's saline content averaged 6,000 parts per million (ppm). This toxic water was discharged into the Gila River until the Reclamation Bureau completed the Wellton-Mohawk Conveyance Channel in the fall of 1961, which deposited the effluent in the Colorado River above Yuma.

Quantity versus Quality

The Reclamation Bureau assumed that the high salinity levels of the wastewater would be diluted by the Colorado, even though by now the Colorado's flow had been seriously diminished by the series of dams that had been constructed upriver, as well as by increased withdrawals from the Colorado by upper basin states. During the 1950s, the Colorado's flow at the international boundary averaged 4.24 million acre-feet a year; in the 1960s, it fell to 1.52 million acre-feet a year.[18] This was just a little more than the amount promised Mexico. The decline in the river's flow meant that there was less freshwater to dilute the newly introduced contaminated wastewater. As a result, in October 1961, the average salinity of water delivered to Morelos Dam in Mexico quickly rose to 1,500 ppm, nearly double the salinity content of water at Imperial Dam: "What had happened was that the Wellton-Mohawk valley had simply transferred its problem downstream to Mexico."[19] Farmers in the Mexicali Valley would now have to deal with the salinity problem, "an additional variable that harmed fields already tottering on the verge of infertility." The contaminated water immediately touched off an ecological crisis, killing crops and damaging farmland in the Mexicali Valley. Mexicali farmers had no recourse but to become more reliant on well water, which was mixed with Colorado River water. The saline water also polluted domestic water supplies on both sides of the border. The City of Yuma, which formerly had taken its domestic water supply from the Colorado, was forced to negotiate an agreement with the Yuma County Water Users Association to receive water from the Yuma Canal, which was connected directly to Imperial Dam.[20]

In November 1961, Mexico lodged a formal complaint against the United States. At the time, the State Department refused to admit that salinity problems in the Mexicali Valley were due to contamination of the Colorado by Wellton-Mohawk farmers, and instead insisted that poor soil and inadequate drainage facilities in the Mexicali Valley were responsible. In order to alleviate the problem, it was suggested that Mexican farmers should improve drainage in their fields and grow more salt-tolerant crops.[21] In any case, the State Department reminded Mexico that the Mexico–U.S. Water Treaty specified only the quantity of water owed Mexico, not the quality. Mexico, at the time the treaty was signed, no doubt realized that its allocation would be met by return flows, since no unused water existed so far downstream, but to insinuate that "Mexico knowingly agreed in the 1944 treaty to accept water unusable for irrigation was 'inconceivable.'"[22]

Several months later, in the winter of 1964, after Lake Powell had begun to fill behind Glen Canyon Dam, salinity levels rose even more, approaching 2,000 ppm at Morelos Dam. Finally, the United States admitted responsibility for the problem, and negotiations for finding a solution began. However, the agreement, referred to as Minute 218 of the International Boundary and Water Commission, was merely a temporary solution.[23] It required the United States to construct a 13-mile drainage bypass from the Wellton-Mohawk Conveyance Channel to a point on the river below Morelos Dam, where the highly saline waters could be dumped in the river either above or below Mexico's last diversion. Mexico could then decide whether it wanted to mix the effluent with better-quality river and well water and use it on Mexicali farms or reject it and discard it into the Gulf of California. Either way, the water delivered to Mexico would still count toward its 1.5 million acre-foot allotment.

Minute 218 originally was a five-year agreement that resulted in moderate improvements in water quality during that period, but it did not permanently solve the salinity crisis. Instead of renegotiating the treaty following its expiration, Mexico agreed to its renewal on an annual basis until a long-term solution could be formulated. Finally, in August 1973, Minute

242 was drafted, which set salinity standards for water deliveries to Mexico at an acceptable level, that is, no more than 115 ppm more saline than the water diverted at Imperial Dam, and guaranteed financial assistance from the United States to help rehabilitate the farms in the Mexicali Valley.[24] To reduce the salinity of the Colorado River, the United States pledged to construct a desalinization plant at Yuma to treat contaminated discharge from Wellton-Mohawk, and to build a bypass drainage channel to carry untreated wastewater directly to the Gulf of California, where it would no longer pollute Mexican water supplies.

Western lawmakers drew on Minute 242 to mesh international with domestic concerns regarding rising salinity levels throughout the Colorado River basin. With each passing year the Colorado's flow was becoming more and more saline. The Colorado's natural salinity load at Lees Ferry, estimated to be 250 ppm, had, by 1972, increased to 606 ppm.[25] The difference was tied to the upper basin states' making increased use of their Colorado River allocations. The Mexican dispute made American irrigators realize that in time they would be facing a similar debilitating problem, and took advantage of Minute 242 to press for additional salinity-control measures throughout the Colorado River basin. In June 1974, Congress gave its overwhelming approval to Minute 242 in the form of a bill known as the Colorado River Basin Salinity Control Act, which would implement measures to control salinity sources in the entire basin.

The Yuma desalinization plant, the second-largest reverse osmosis plant in the world, originally was due to go online in 1981, but a variety of funding and design problems pushed back its completion date to 1992. It was constructed at a cost of $250 million, but after a brief period of operation, it was mothballed when heavy flooding along the Gila River in early 1993 destroyed parts of the canal that delivered water to the plant from Wellton-Mohawk. Afterward, it remained shut down because it was deemed too expensive to run, particularly since a series of wet years made it more efficient to meet the Mexican treaty obligations with water from Lake Mead. Hence, so far, the Yuma desalter has served as a costly insurance policy to prevent violations of Minute 242. Meanwhile, the 59-mile bypass canal, which carries untreated drainage directly to the Gulf of California, has been a much cheaper alternative, and has relieved pressure to use the plant.

So long as the river was at normal flow, the desalter simply was not needed, but the current multiyear drought in the Colorado River basin has brought the plant to the forefront: Currently, 100,000 acre-feet of water are

being released from a much reduced Lake Mead to help fulfill treaty obligations with Mexico.[26] If the plant were to be operated, desalted water could
substitute for reservoir water, thereby helping to alleviate possible water
shortages in the lower basin. In fact, because of the ongoing drought, after
being shut down for fifteen years, the Yuma desalting plant was operated
in a demonstration run at 10 percent capacity from March 31 to May 31,
2007. More than 4,000 acre-feet of water were returned to the river at less
cost than projected.[27] It was Arizona water managers, who will suffer water cuts first in the event of a shortage on the Colorado, who called for the
plant to be turned on, not necessarily for the purpose of potentially meeting
treaty obligations with Mexico but for the possibility of supplying municipal-
quality water to growing communities around Yuma.[28] Still, voluntary sales
and transfers of water from willing sellers would be a cheaper alternative to
operating the plant.[29]

Meanwhile, the untreated wastewater from the Wellton-Mohawk Irrigation and Drainage District being discharged into the Gulf of California
has given rise to something totally unexpected—the creation of Ciénega de
Santa Clara, the largest open-water wetland in the Colorado delta. Although
Ciénega de Santa Clara, which encompasses more than 40,000 acres of lush
wetlands, is located slightly northeast of the mouth of the Colorado and is
physically disconnected from what remains of the river's natural delta, it has
grown to serve a critical ecological role in the delta ecosystem, and today it
is protected by Mexico through its designation as a biosphere reserve.[30] This
emerging wetland is now approximately the size of the present-day natural
delta, which once covered nearly 1.8 million acres. The natural delta's shrink-
age is due to the construction of both Hoover and Morelos dams, which severely restricted river flows to the delta region, and to the completion of Glen
Canyon Dam in 1964, which virtually eliminated river flows to the delta. If
the Yuma desalter were to be placed in operation to meet treaty obligations
with Mexico, the environmental costs would be substantial, for the Ciénega
de Santa Clara would dry up with the elimination of its saline water supply.
Saving the wetlands of the Colorado River delta region is an environmental
issue that no doubt will gain significant prominence in the future.[31]

The "4.4 Plan"

As states located in the Colorado River basin began to use increasing shares
of their own allotments from the river, troubles other than those related to
increased salinity levels began to emerge. Most notably, California, with its

two Colorado River conduits—the All-American Canal and the Colorado River Aqueduct—was using far more water than its allocation of 4.4 million acre-feet per year stipulated by the Colorado River Compact, and reinforced by the 1963 Supreme Court ruling. As long as other basin states were not drawing off their full allocations, supplemental water allotments amounting to more than 600,000 acre-feet per year were consistently available for California's use. Much of this surplus is destined for metropolitan Southern California.

In 1968, the Colorado River Basin Project Act was approved by Congress, which authorized the construction of the Central Arizona Project. The Central Arizona Project was designed to carry Arizona's remaining share of Colorado River water to the state's water-deficient areas, notably Maricopa (Phoenix), Pinal, and Pima (Tucson) counties. As early as 1944 the Bureau of Reclamation had formulated various preliminary plans for a Central Arizona Project, but California argued that the project could not be approved until disagreements involving the two states' annual allotments from the Colorado were determined. To gain California's support, Arizona had to agree that California could take its full allotment of 4.4 million acre-feet before any water was delivered to the Central Arizona Project. Arizona, therefore, would be the first to suffer water cutbacks during times of shortage. This prompted Arizona in 1996 to create the Arizona Water Bank that enables the state to utilize, via storage in aquifers, its full share of Colorado River water. The creation of the water bank prevents California from continuing to rely on Arizona's unused water while it simultaneously improves the reliability of central Arizona's water supply.[32]

Once the Central Arizona Project was completed, and Arizona and other basin states began to use more of their legitimate share of the Colorado's flow, they demanded that the federal government require California to rein in its overreliance on the river. In December 2000, Interior Secretary Bruce Babbitt interceded in the conflict and fashioned an agreement that would compel California to live within its means. The accord, known as the "4.4 Plan," required California to begin to wean itself off "surplus" Colorado River water by January 1, 2003, and to continue to cut back during a fifteen-year transition period after which it would receive no more than its firm allocation of 4.4 million acre-feet per year. With a decade and a half to achieve this goal, this process was referred to as the "soft landing" approach to reducing California's overdependence on the Colorado. Four water agencies—the Metropolitan Water District of Southern California (MWD), the San Diego

County Water Authority (SDCWA), the Imperial Irrigation District (IID), and the Coachella Valley Water District (CVWD)—were given two years, until December 31, 2002, to determine how this would be achieved. Agreement would not be easy, since 4.4 million acre-feet was significantly less than the 5.2 million acre-feet of water that California was diverting from the river, not to mention future anticipated water needs of the four participating water agencies.

Because of the legally recognized prior appropriations made more than a century ago, approximately 70 percent of California's Colorado River allocation goes to agricultural uses in Imperial and Riverside counties, which include the irrigation districts serving the Imperial, Coachella, and Palo Verde valleys and the California side of the Yuma Project.[33] The Imperial Irrigation District alone, which provides water for about 150,000 residents and 500,000 acres of farmland, is by far the single largest consumer of Colorado River water, with an enormous annual entitlement of approximately 3.1 million acre-feet, more than any single state, excluding California.[34] Although the Imperial Irrigation District would face some cutbacks during a severe shortage, it would still be able to continue to irrigate even when suburban Los Angeles and San Diego receive no water. The reason is that only about 13 percent of the state's firm allocation, or 550,000 acre-feet, is available for use by the Metropolitan Water District of Southern California, which serves 18 million people in six counties in coastal Southern California, including the San Diego County Water Authority. Both the MWD and the SDCWA wholesale water to their client agencies. This amount is not nearly enough to meet their clients' needs, and is less than half the amount of water the Colorado River Aqueduct was designed to carry. By supplementing the 550,000 acre-feet with 662,000 acre-feet of "surplus" water each year (water that the other basin states were not using), the Colorado River Aqueduct for the past two decades has operated at near capacity levels. What this means is that the MWD and the SDCWA would have to find substitute sources from the Colorado to keep the aqueduct full as California's supplemental allocation gradually dries up.

Except for the implementation of water conservation measures among residents of Southern California, the obvious place to look to help offset the impending shortages are California's agricultural districts, specifically the Imperial Irrigation District, that use a disproportionate share of California's Colorado River allocation. In 1988, after the Imperial Irrigation District was accused by the state of wasting water, Imperial agreed to transfer

100,000 acre-feet of conserved water to the MWD[35] The conserved water will be purchased by the MWD's member water districts, which will fund the various conservation measures. More recently, a tentative seventy-five-year agreement was reached between the IID and the SDCWA whereby the former would sell up to 200,000 acre-feet of water per year to the latter by the year 2020.[36] Until now, San Diego, because of the scarcity of ground-water reserves, has been almost totally reliant on the MWD to supply it water from the Colorado River; in a severe water shortage, San Diego's portion would be the first to be cut back. In fact, the Imperial–San Diego sale was a key to the state's ability to live within its firm allocation of 4.4 million acre-feet, and it was the pivotal provision in the water agreement that was to be consummated by the four participating water agencies by December 31, 2002.

As the deadline approached, in early December 2002, the Imperial Irrigation District began to have serious reservations concerning the San Diego water transfer, and rejected the proposed sale.[37] The federal government had repeatedly emphasized that if a deal was not reached by year's end, then the "hard landing" option would be implemented—that is, California's supplemental allocation would be cut off immediately on January 1, 2003. The hard landing would primarily impact coastal Southern California, since it was most reliant on the "surplus" water, but the Interior Secretary also threatened to cut Imperial's 3.1 million acre-foot allocation by up to 10 percent in order to partially compensate Southern California for its loss.

To justify this action, the federal government invoked a 1979 decree that supplemented the 1963 Supreme Court decision that divided the Colorado's flow among the lower basin states, a ruling that set limits on how much water farmers could use per acre. Imperial Valley farmers, who view Colorado River water as their birthright, were consistently above those per-acre limits, prompting the Interior Department to accuse valley farmers of wasting water, thereby ordering that their mammoth allocation be reduced.[38] The Imperial Irrigation District refused to be intimidated, however, and immediately filed a countersuit against the government. In March 2003, a U.S. district judge ordered the Imperial cuts restored, but he also instructed the Bureau of Reclamation to conduct a "beneficial use" review of the Imperial Irrigation District's irrigation practices.

The Imperial Irrigation District backed out of the water agreement because of two principal concerns. Imperial farmers had understood that the water to be sold to San Diego would come through the implementation of conservation measures, not through letting fields go fallow. Although improvements to the canals are planned, as are measures to capture unused water, the agreement called for immediate fallowing, a temporary measure lasting for as much as fifteen years while farmers install conservation devices in their fields.[39] With much of the West in the grip of a prolonged drought, a speedier solution to the impending water shortages in Southern California was required. Imperial Valley interests feared that widespread fallowing would undermine the region's $1 billion agricultural economy, causing negative multiplier effects resulting from increased unemployment, business bankruptcies, and the loss of tax revenue for local governments and schools. Most economists, however, believe the negative impacts on the local economy would not be significant, assuming the in-lieu payments that participating farmers would receive would be spent locally.[40]

The chief deal breaker, however, involved the probable environmental effects of fallowing on the Salton Sea. State and federal law requires that parties to a water deal repair any environmental damage resulting from such an agreement. Currently, much of the excess water from farmers' fields and the unused water flowing through the distribution canals end up in the Salton Sea; that is, the Salton Sea survives today on agricultural runoff amounting to approximately one-third of the water that is delivered to the Imperial Valley via the All-American Canal. It is because of this excessive runoff that Imperial is widely viewed as wasteful.

Nevertheless, the Imperial Irrigation District feared that it would be billed hundreds of millions of dollars for environmental restoration projects for damage done to the Salton Sea after the water transfer to San Diego commenced. The Salton Sea, once a popular recreation spot in California, and major stopover point for millions of migratory birds, has suffered a sharp decline in recent years; it is plagued by odors, increasing salinity, and occasional massive die-offs of fish and birds. The San Diego sale would decrease the amount of runoff, making the Salton Sea smaller and even more foul-smelling than it currently is, and more lethal to birds and fish. As the level of the sea falls over time, large areas of bottomlands that contain salts and other environmentally harmful chemicals will be exposed. Toxic "dust-salt" plumes, similar to those that pollute the Owens Valley that rise from the dry bed of Owens Lake under windy conditions, could become common occurrences in the Imperial and Coachella valleys in the future, regions that already suffer from significant air pollution. At the time that the San Diego sale was tentatively agreed to, the participating water agencies seemed unwilling to invest large sums for the environmental restoration of the land-locked sea, and, hence, the original deal fell through.

After the failure of the agreement, the "hard landing" was immediately implemented in California. However, the sharp reduction in water supply spurred new negotiations, and ten months later, in October 2003, an accord, known as the Quantification Settlement Agreement, was signed by the four water agencies involved in trying to wean California off its supplemental Colorado River allocations. In accordance with the original plan, the Impe-rial Irrigation District will sell up to 200,000 acre-feet of water per year for at least seventy-five years to the SDCWA This represents the nation's biggest shift ever of water from farms to cities. The MWD will no longer be the sole water provider to San Diego, its largest customer, but has agreed to allow its aqueduct to be used to carry Imperial water to San Diego. The accord also allows an unspecified additional amount of water to be sold to the state, which in turn will resell it at a profit to the MWD. This will reduce Southern California's reliance on having to transfer water from northern California farms. A $300 million fund meant to stop the environmental decline of the Salton Sea will be created, and new legislation will set a dollar limit on how much the participating agencies could be billed for Salton Sea restoration. The State of California will assume responsibility for environmental damage above the cap set on the amount local agencies will have to pay.[41]

Even with the complex water agreement now signed, the road ahead will not be a smooth one. Although California once again is eligible for surplus allocations from the river for more than a decade, if the current drought per-sists, it is unclear whether there will be enough water in Lake Mead to send surplus water to Southern California. According to the U.S. Geological Sur-vey, the period between 1999 and 2004 was the driest period in the ninety-eight years of recorded history of the Colorado. Although 2005 turned out to be a wet year, the drought seems to have returned, and, for the first time since Hoover Dam was built to control the river's flow, the basin states are preparing for the possibility of water shortages.[42] A prolongation of the drought could mean that a "hard landing" for Southern California may still be unavoidable. In fact, there are signs that the "perfect Southern California drought" might be setting up whereby every place that supplies water to the region's eighteen million inhabitants is unusually dry.[43] Not only is the Colo-rado River basin suffering from one of the worst droughts in a long time (the river supplies 65 percent of the water distributed in Southern Califor-nia), but rainfall in the winter of 2007 in Los Angeles was the lowest ever recorded, and the Sierra snowpack, also vital to Southern California's water imports, is at its lowest level in nearly two decades.

Crisis and Conflict

It has become clear that the river's outlook today is even worse than federal "worst-case" scenario predictions of the mid-1990s.[44] It is now forecast that global warming will likely mean even less water for the Colorado in the future. A committee of the National Research Council recently reported that the twentieth century saw a trend of increasing mean temperatures across the Colorado River basin that has continued into the early twenty-first century, and there is no evidence that this warming trend will dissipate in the coming decades. Although it is uncertain how rising temperatures will affect precipitation patterns, higher regional temperatures already are having an impact on snowpack in the Colorado basin, the lifeblood of the river. Rising future temperatures will result in less snowpack accumulation compared to the present, and will shift the timing of peak spring snowmelt to earlier in the year. Higher average temperatures will also contribute to increases in water demands, especially during summer, as well as increased evaporative losses from snowpack, surface reservoirs, irrigated land, and vegetated surfaces.[45] Some scholars believe that climate change will turn future droughts into megadroughts. According to Glen MacDonald, "Given the climate warming of the past decades and the projected warming over the next century, it is possible that we are already in one."[46] Even absent the probability of global warming, tree-ring reconstructions that extend the hydrological record back centuries indicate that severe multiyear and decadal-scale droughts are a defining feature of the upper Colorado River basin that supplies most of the river's water.[47] In light of the prehistoric record, "the recent Colorado Basin drought is not exceptional and dry conditions could conceivably persist into future years."[48]

But the river's plight is not just a result of the current and possible prolonged dryness. It is also due to explosive growth of the lower basin states. Since the Colorado River Compact was signed, large population increases have been experienced in many of the basin's major cities such as Phoenix, Tucson, and Las Vegas, as well as several cities on the basin's periphery that depend on Colorado River water, including Los Angeles, San Diego, Albuquerque, and Denver. Since the 1920s, the population of Phoenix, for example, has grown by forty-seven times and Las Vegas by two hundred times. Even though some innovative urban water use and conservation programs have led to reductions in per-capita use, this unprecedented growth has generated a 410 percent increase in domestic water use in the Southwest since 1950.[49] Population growth rates and future projections are on a sharply increasing trajectory in the western United States, and they point to sizable and growing water demands for the foreseeable future. The National Research Council warns that "technological and conservation options for augmenting or extending water supplies—although useful and necessary—in the long run will not constitute a panacea for coping with the reality that water supplies in the Colorado River basin are limited and that demand is inexorably rising."[50]

Aggravating the crisis is the 1922 miscalculation of the Colorado's discharge when hydrologists overestimated the river's average flow and locked the number into the multistate Colorado River Compact. Instead of the original estimate of 16.4 million acre-feet, a figure based on river gauging during an unusually wet period between 1905 and 1922 (tree-ring reconstructions indicate that this was one of the wettest periods in the past five hundred years), the long-term average flow since 1905 is about 15 million acre-feet per year, whereas tree-ring analyses that extend the record back to the 1500s suggest that the average flow is significantly lower than both of the above figures.[51] Consequently, under normal circumstances, more of the Colorado has been allocated to the seven basin states and Mexico than the river can be expected to provide in sustained fashion.

The reservoirs in the Colorado basin, with a combined storage capacity of about 60 million acre-feet, can hold four years of supply using an average annual flow of 15 million acre-feet.[52] They are designed to allow the system to provide adequate water to users during times of drought, but the current drought is straining the capacity of those reservoirs to mitigate the low flows, and if the drought continues it could soon overwhelm the buffering capacity of the reservoirs. Annual inflow to Lake Powell provides a useful barometer of drought conditions in the Colorado River Basin. In the 2000 water year (October 1999 through September 2000) inflow was 62 percent of normal, 2001 saw 59 percent of normal, 2002 was only 25 percent of normal (the lowest ever recorded since Lake Powell began filling in 1963), 2003 rebounded somewhat to 51 percent of normal, while 2004 was 49 percent of normal. The following year, 2005, was relatively wet in the basin, resulting in inflow 105 percent of normal, but in 2006 inflow once again fell below normal to 73 percent, and the inflow for the 2007 water year was 68 percent of normal.[53] Drought conditions again eased somewhat during the 2008 water year with an inflow of 106 percent. In 1999, the two megareservoirs on the Colorado, Lake Powell and Lake Mead, with a combined storage capacity of 55 million acre-feet, were virtually full; current reservoir storage in Lake Powell is 61 percent of capacity, while storage in Lake Mead is 46 percent of capacity (fig. 10.2). Yet even after seven out of nine below-average years of precipitation and inflows (2000–2008), the Colorado River storage system still holds roughly two years of average annual Colorado River flows. However, because of rapid population growth and water demand, refilling the reservoirs will be much more difficult than in the past. It is now estimated that it would take fifteen years of normal precipitation and runoff conditions to refill these two massive reservoirs.[54]

Preparing for the possibility of water shortages for the first time since the Colorado River Compact was signed, the seven basin states, and the U.S. Bureau of Reclamation (representing the Interior Department), are considering strategies that would establish procedures for what to do if the river cannot meet the demand for water, a prospect that some experts predict could occur in about five years.[55] "Whereas basin states once negotiated ways to fairly and equitably share Colorado River water (or at least achieve as close an approximation as possible) they are now discussing ways to fairly and equitably share water shortages."[56] In 2005, the Interior Department put the seven states on notice that if they did not devise a plan to deal with shortage among themselves by December 2007, federal officials would step in and impose water restrictions along the river when necessary.[57] In February 2006, the seven basin states began to develop shortage guidelines and management strategies under low-reservoir conditions.[58] In December 2007, federal officials signed the "Colorado River Interim Guidelines for Lower Basin Shortages and the Coordinated Operations for Lake Powell and Lake Mead," a pact that includes a bundle of agreements among the basin states on how to allocate water if the river runs short.[59] It establishes criteria for the Interior Department to declare a shortage on the river, which would occur when the system is unable to produce the 7.5 million acre-feet of water allocated to lower basin states, and how it would manage the two major reservoirs in times of water shortage, and it spells out how the three lower basin states will share the impact of water shortages.

The new accord also outlines measures to encourage conservation. One component of the agreement promoting conservation is the "Drop 2" project. Drop 2 is a reservoir that will be located on the All-American Canal about twenty-five miles west of Yuma, the purpose of which is to salvage

FIG. 10.2. Lake Mead, November 2004. At this time, after five years of severe drought, Lake Mead's storage was 46 percent of capacity. In late August 2008, the reservoir's storage stood at the same level. The top of the white rim reflects the high-water mark in the reservoir, reached most recently in 1999. The height of the white rim illustrates the drawdown in storage reserves. (Photo by the author)

water from a now overallocated river. The U.S. Bureau of Reclamation will operate Drop 2, but it is being funded by Nevada, Arizona, and California as a cooperative venture. It therefore reflects current political and hydro-logical realities far different from those that prevailed during previous eras along the river, and serves as an example of states working together to solve Colorado River issues.[60] The new reservoir is designed to capture inadvertent overdeliveries and flash floods that otherwise would escape unused into Mexico. Most of this water is lost when agricultural clients on the lower end of the river, because of changing weather conditions or because of irrigation equipment malfunction, for example, fail to accept the water they ordered. Since it normally takes about one week for water released from Lake Mead to reach the farms on the lower Colorado, conditions for farmers might change in the interim. Drop 2, then, would provide temporary storage until the water is returned to the system for use. Since the water level in Lake Mead is the prime deciding factor determining when a shortage on the lower Colorado River is declared, the reservoir is intended to slow water declines in Lake Mead, thereby delaying the proclamation of a shortage. Currently, these so-called nonstorable flows are not counted against Mexico's legal allocation of 1.5 million acre-feet. Farmers in Mexico utilize some of this water, but much of the undedicated flow helps support what is left of the habitat-rich Colorado River delta ecosystem. Matt Jenkins suggests that "the biggest costs with this new obsession with efficiency could ultimately accrue to the very place that bore the brunt of the first round of development: the foundering ecosystem of the Colorado River Delta."[61]

If the Colorado River goes into crisis mode, the new guidelines for lower basin shortages are expected to forestall litigation over water rights that could dismantle the so-called law of the river, a mix of federal and state statutes, interstate compacts, congressional acts, and legal decisions, which is considered the ultimate authority in allocating and using Colorado River water. Some, however, believe an overhaul of the law of the river would be a good thing. According to Daniel McCool, "The law of the river is hopelessly, irretrievably obsolete, designed on a hydrological fallacy, around an agrarian West that no longer exists."[62]

Given the oversubscribed nature of the Colorado River, one cannot argue with the hydrological fallacy that currently prevails, but some might take issue with the statement that an agrarian West no longer exists. Granted, there are far more urban dwellers than rural residents in the West today, but the rural folks still hold the lion's share of water rights in the region from which they generate substantial wealth from a wide variety of agricultural products. The Colorado River Compact and much of the law of the river were framed during an era when water for irrigation was of paramount concern, and roughly 80 percent of the Colorado River basin water is still allocated to the agricultural sector. Hence, farming and ranching activities remain imbedded in the western landscape in spite of rampant urban growth. But as we have seen, in the Imperial Valley the agrarian influence is waning, and in fact, "agricultural water appears to constitute the most important, and perhaps final, large reservoir of water available for urban use in the arid U.S. West." There are, however, both direct and third-party barriers to transferring water to municipalities from agricultural users, such as reduced food-production capability, reduced agricultural return flows that support riparian ecosystems, lost business suffered by local merchants as a result of reductions in irrigated cropland, and limited physical facilities for storing and rerouting water among willing buyers and sellers.[63]

It has now become apparent that the Colorado River under normal flow conditions is insufficient to meet current water allocations if all seven basin states, and Mexico, were to withdraw their full allotments. It is also clear from both an ecological and an economic standpoint that we will continue to face significant challenges if irrigated agriculture is to remain a sustainable activity in the face of unbridled urban growth in the region. In fact, the Interior Department, in the context of the Quantification Settlement Agreement, has gone on record, in California at least, of favoring urban users over agricultural users of Colorado River water. The recent agreement has allowed the cities to advance significantly up the California priority system, and represents a fundamental shift in the balance of power on the river.[64] According to the National Research Council, "The combination of limited Colorado River water supplies, rapidly increasing populations and water demands, warmer regional temperatures, and the specter of recurrent drought point to a future in which the potential for conflict among existing and prospective new users will prove endemic."[65] When the Colorado began to be diverted for agricultural purposes at Yuma little more than a century ago, nobody could have imagined the conflicts and controversies that would eventually develop resulting from the growing demand for water from the river's overtapped and rapidly diminishing supply. Based on the events that have unfolded along the lower Colorado River, it is difficult to argue with Jenkins's observation that "we have not civilized the West so much as savaged it."[66]

Notes

1. THE FINAL FRONTIER

1. *U.S. Statutes at Large* 32 (1902): pt. 1, 388.

2. See, for example, Mark Fiege, *Irrigated Eden: The Making of an Agricultural Landscape in the American West;* Nancy Langston, *Where Land and Water Meet: A Western Landscape Transformed,* 1–62; and Robert A. Sauder, *The Lost Frontier: Water Diversion in the Growth and Destruction of Owens Valley Agriculture.* Fiege discusses how an irrigated landscape came to be in the Snake River valley of southern Idaho, Langston describes how ranchers and homesteaders transformed the wetlands ecosystem of the Malheur Lake Basin in their attempt to settle this dry region of Oregon, and Sauder focuses on the development and destruction of irrigated agriculture in the Owens Valley of California.

3. For a general overview of federal irrigation policy and the problems related to it, see, for example, Augustus Griffen, "Land Settlement of Irrigation Projects, With Discussion"; Alvin Johnson, "Economic Aspects of Certain Reclamation Projects"; Dorothy Lampen, *Economic and Social Aspects of Federal Reclamation;* Elwood Mead, "Present Policy of the United States Bureau of Reclamation Regarding Land Settlement, With Discussion"; F. L. Tomlinson, "Land Reclamation and Settlement in the United States"; U.S. Congress, Senate, *Federal Reclamation by Irrigation;* Department of the Interior, U.S. Geological Survey, *First Annual Report of the Reclamation Service, From June 17 to December 1, 1902,* 15–75; Department of the Interior, *Fourteenth Annual Report of the Reclamation Service, 1914–1915,* 1–25; Department of the Interior, Bureau of Reclamation, *Twenty-third Annual Report of the Bureau of Reclamation,* 1–10; and John A. Widstoe, "History and Problems of Irrigation Development in the West, With Discussion."

4. Donald Pisani, *Water and the American Government: The Reclamation Bureau, National Water Policy, and the West, 1902–1935,* 154; Donald Pisani, "Irrigation, Water Rights, and the Betrayal of Indian Allotment," 172.

5. Pisani, "Irrigation, Water Rights, and Betrayal," 158, 171; Daniel McCool, *Command of the Waters: Iron Triangles, Federal Water Development, and Indian Water,* 13.

6. Pisani, *Water and the American Government,* 154.

7. McCool, *Command of the Waters,* 112.

8. Pisani, "Irrigation, Water Rights, and Betrayal," 158.

9. Ibid., 160–61.

10. For background on how irrigation impacted other Indian groups, see, for example, Lloyd Burton, *American Indian Water Rights and the Limits of Law;* Martha C. Knack and Omer C. Stewart, *As Long As the River Shall Run: An Ethnohistory of Pyramid Lake Indian Reservation;* Barbara Leibhardt, "Allotment Policy in an Incongruous Legal System: The Yakima Indian Nation As a Case Study, 1887–1934"; David Rich Lewis, *Neither Wolf nor Dog: American Indians, Environment, and Agrarian Change;* Thomas R. McGuire, William B. Lord, and Mary G. Wallace, eds., *Indian Water Rights in the New West;* Pisani, "Irrigation, Water Rights, and Betrayal"; Pisani, *Water and the American Government,* 181–201; and John Shurts, *Indian Reserved Water Rights: The Winters Doctrine in Its Social and Legal Context, 1880s–1930s.*

11. Pisani, *Water and the American Government,* 200.

12. Robert L. Bee, *Crosscurrents Along the Colorado: The Impact of Government Policy on the Quechan Indians,* 160.

13. Emily Greenwald, *Reconfiguring the Reservation: The Nez Perces, Jicarilla Apaches, and the Dawes Act,* 11, 153.

14. Lewis, *Neither Wolf nor Dog,* 3.

15. Flood recession agriculture dominated the domesticated landscapes along the lower courses of many rivers in the far southwestern corner of North America from the lower Yaqui River in southern Sonora, Mexico, to the lower Colorado River in Northwest Mexico and the Arizona-California border area, although evidence of its precise distribution is limited because this form of irrigation did not require any permanent landscape modifications that would provide archaeological relics for study (William E. Doolittle, *Cultivated Landscapes of Native North America,* 414–15, 427).

16. Jonathan B. Mabry and David A. Cleveland, "The Relevance of Indigenous Irrigation: A Comparative Analysis of Sustainability," 227; Doolittle, *Cultivated Landscapes,* 4.

17. Michael C. Robinson, *Water for the West: The Bureau of Reclamation, 1902–1977,* 2.

18. Donald Worster, *Rivers of Empire: Water, Aridity, and the Growth of the American West,* 35.

19. Some present-day Native Americans in the Southwest, such as the Tohono O'odham (Papago) and the Pima, have learned from their ancestors' mistakes, surviving in the desert "with a lighter touch" aimed at achieving a secure coexistence and a thrifty subsistence. Today these groups practice flood recession farming by locating their fields at the mouths of the arroyos of small streams where the waters of the floods slow down, spread out, and soak into the soil. See Wendell Berry, *The Gift of the Good Land: Further Essays Cultural and Agricultural,* 54; and Worster, *Rivers of Empire,* 33–35.

20. Fiege, *Irrigated Eden,* 12, 24.

21. Mabry and Cleveland, "Relevance of Indigenous Irrigation," 227; Donald Worster, "Thinking Like a River," 60.

22. Mabry and Cleveland, "Relevance of Indigenous Irrigation," 228–29. It can be argued, however, that the beneficiaries of irrigation include not just farmers but

also, indirectly, many others, including firms supplying farm inputs, firms processing and marketing farm outputs, and, of course, consumers of those outputs. See Ian Carruthers and Colin Clark, *The Economics of Irrigation*, 187.

23. Berry, *Gift of the Good Land*, 63; Worster, *Rivers of Empire*, 51; Berry, *Gift of the Good Land*, 66.

24. Aldo Leopold, *A Sand County Almanac—With Essays on Conservation From Round River*, 159.

25. Department of the Interior, U.S. Geological Survey, *First Annual Report of the Reclamation Service*. In the 1880s nearly 4 million acres were brought under irrigation, and approximately 3.5 million irrigated acres were added in the 1890s.

26. John T. Ganoe, "The Desert Land Act in Operation, 1877–1891"; Robert A. Sauder, "Patenting an Arid Frontier: Use and Abuse of the Public Land Laws in Owens Valley, California."

27. Robinson, *Water for the West*, 8–10.

28. Congress in 1887 instructed the director of the Geological Survey to consider the question of federal irrigation. Appropriations were subsequently made to examine the extent to which the arid regions could be redeemed by irrigation. Results of the investigation were published in the tenth, eleventh, twelfth, and thirteenth annual reports of the Geological Survey. This series of irrigation reports came to be known as the Powell Irrigation Survey.

29. Robinson, *Water for the West*, 12.

30. U.S. Congress, Senate, *Report of the Special Committee of the United States Senate on the Irrigation and Reclamation of Arid Lands*.

31. Robinson, *Water for the West*, 13.

32. John T. Ganoe, "The Origin of a National Reclamation Policy," 35; Robinson, *Water for the West*, 11.

33. Worster, *Rivers of Empire*, 140. See pp. 138–40 for more detail on Powell's recommendations.

34. Robinson, *Water for the West*, 9.

35. U.S. Congress, Senate, *Federal Reclamation by Irrigation*, 35.

36. Norris Hundley Jr., *Water and the West: The Colorado River Compact and the Politics of Water in the American West*, 9–10.

37. Donald Pisani, "Federal Reclamation and the American West in the Twentieth Century," 393; Robinson, *Water for the West*, 18; Worster, *Rivers of Empire*, 130.

38. The Bufford-Trenton Project was approved in 1904, but poor soil quality caused landowners to lose interest, and the project was discontinued. Robinson states that twenty-five projects were approved through 1907, but several were auxiliary to the projects named herein. In 1922, according to Robinson, twenty-two projects were active, with several having been abandoned due to engineering and financial problems (*Water for the West*, 20).

39. Quoted in Department of the Interior, U.S. Geological Survey, *First Annual Report of the Reclamation Service*, 44.

40. Pisani, *Water and the American Government,* 30; Pisani, "Federal Reclamation and the American West," 397–98.

41. Robinson, *Water for the West,* 19; Pisani, *Water and the American Government,* 114.

42. Department of the Interior, U.S. Geological Survey, *First Annual Report of the Reclamation Service,* 43.

43. Pisani, *Water and the American Government,* 30; Robinson, *Water for the West,* 38.

44. Marc Reisner, *Cadillac Desert: The American West and Its Disappearing Water,* 118.

45. Mead, "Present Policy," 732; Robinson, *Water for the West,* 45. Three of the early projects had no public land at all.

46. William Wyckoff, "Understanding Western Places: The Historical Geographer's View," 44.

2. QUECHAN LAND

1. The mountain is located north of present-day Needles, California, at the southern tip of Nevada and is known officially as Newberry Mountain, but is also called "Ghost Mountain" by non-Indians. See Jack D. Forbes, *Warriors of the Colorado: The Yumas of the Quechan Nation and Their Neighbors,* 4; C. Daryll Forde, *Ethnography of the Yuma Indians,* 88; and William deBuys and Joan Myers, *Salt Dreams: Land and Water in Low-Down California,* 260–61.

2. Forde, *Ethnography of the Yuma Indians,* 89.

3. Raymond W. Stanley, "Political Geography of the Yuma Border District," 1:50–52.

4. A. L. Kroeber, *Cultural and Natural Areas of Native North America,* 42.

5. Stanley, "Political Geography," 1:56, 53.

6. John P. Harrington, "A Yuma Account of Origins," 325; Stanley, "Political Geography," 1:55.

7. Stanley, "Political Geography," 1:55.

8. Edward F. Castetter and Willis H. Bell, *Yuman Irrigation Agriculture: Primitive Subsistence on the Lower Colorado and Gila Rivers,* 39; Fred B. Kniffen, *Lower California Studies, III: The Primitive Cultural Landscape of the Colorado Delta,* 50.

9. Malcolm J. Rodgers, "An Outline of Yuman Prehistory," 168–69.

10. Harrington, "A Yuma Account of Origins," 342; Kroeber, *Cultural and Natural Areas,* 42.

11. Homer Aschmann, "The Head of the Colorado Delta," 233; Fred B. Kniffen, *Lower California Studies, IV: The Natural Landscape of the Colorado Delta,* 205.

12. R. H. Forbes, *The River-Irrigating Waters of Arizona: Their Character and Effects,* 206.

13. Kniffen, *Lower California Studies, III,* 43.

14. S. P. Heintzelman, "Headquarters, Fort Yuma, California, July 15, 1853," 36.

15. Aschmann, "Head of the Colorado Delta," 251.

16. For a more detailed account of flood recession farming along the lower Colorado River, see Doolittle, *Cultivated Landscapes*, 422–26.

17. Aschmann, "Head of the Colorado Delta," 251.

18. Eugene J. Trippel, "The Yuma Indians," 574.

19. Aschmann, "Head of the Colorado Delta," 252.

20. Castetter and Bell, *Yuman Irrigation Agriculture*, 39, 52.

21. Kniffen, *Lower California Studies, III*, 49.

22. J. Forbes, *Warriors of the Colorado*, 84.

23. Aschmann, "Head of the Colorado Delta," 248.

24. It was not until Garcés and Anza traveled through this area that the ecology of Yuman flood recession farming became known to people of European ancestry. See Doolittle, *Cultivated Landscapes*, 419.

25. Roger Dunbier, *The Sonoran Desert: Its Geography, Economy, and People*, 145.

26. The name Camino del Diablo applied to the particularly hazardous 150-mile stretch between Sonoita and the Gila River at present-day Yuma. The first recorded journey along this route was made by Father Kino in 1699. See Godfrey Sykes, "The Camino del Diablo: With Notes on a Journey in 1925," 62–64.

27. Hubert Howe Bancroft, *The Works of Hubert Howe Bancroft*, 396.

28. Ibid.

29. Stanley, "Political Geography," 1:68.

30. J. Forbes, *Warriors of the Colorado*, 220, 256.

31. *United States v. Coe*, 747.

32. Clifford E. Trafzer, *Yuma: Frontier Crossing of the Far Southwest*, 51.

33. J. Forbes, *Warriors of the Colorado*, 298.

34. Ibid., 303.

35. Ibid., 319.

36. Trafzer, *Yuma*, 57.

37. Ibid., 59.

38. Heintzelman, "Headquarters, Fort Yuma, California," 45, 50.

39. Ibid., 50.

40. Trafzer, *Yuma*, 82–83.

41. Ibid., 112.

42. Trippel, "The Yuma Indians," 9.

43. Executive Order, July 6, 1883; Executive Order, January 9, 1884; *Quechan Tribe of Fort Yuma Reservation, California*, 55.

44. Eugene J. Trippel, "Report of the Citizens' Executive Committee on the Irrigation and Reclamation of Arid Lands," 27.

45. Trippel, "The Yuma Indians," 562.

46. Aschmann, "Head of the Colorado Delta," 253.

3. EARLY IRRIGATION VENTURES

1. Clyde P. Ross, *The Lower Gila Region, Arizona: A Geographic, Geologic, and Hydrologic Reconnaissance With a Guide to Desert Watering Places*, 97.

2. Richard J. Hinton, *The Hand-Book to Arizona: Its Resources, Towns, Mines, Ruins, and Scenery,* 278–79.

3. Cora Savant Nicholas, "The History of Yuma Valley and Mesa With Special Emphasis on the City of Yuma, Arizona," 72.

4. Ibid., 73; Ross, *Lower Gila Region,* 97.

5. Trippel, "Report of the Citizens' Executive Committee," 9, 14–15; Trippel, "The Yuma Indians," 10.

6. Trippel, "Report of the Citizens' Executive Committee," 21–24.

7. Robert G. Schonfeld, "The Early Development of California's Imperial Valley," 285; C. E. Grunsky, "The Lower Colorado River and the Salton Basin," 18.

8. Schonfeld, "Early Development of Imperial Valley," 286.

9. U.S. Congress, Senate, "Letter From the Secretary of the Interior, Transmitting a Copy of an Agreement With the Yuma Indians, With a Report From the Commissioner of Indian Affairs and Accompanying Papers," 28.

10. U.S. Congress, House, "Agreement With Yuma Indians in California."

11. Ibid., 1.

12. U.S. Congress, Senate, "Letter From the Secretary," 16–17.

13. Nicholas, "History of Yuma Valley and Mesa," 95, 96.

14. Ibid., 96.

15. U.S. Congress, Senate, "Letter From the Secretary," 16.

16. Ibid., 15.

17. *Quechan Tribe of Fort Yuma Reservation,* 90.

18. U.S. Congress, Senate, "Letter From the Secretary," 7.

19. Ibid., 10.

20. Ibid., 24, 26.

21. *Quechan Tribe of Fort Yuma Reservation,* 91.

22. Ibid., 87.

23. Grunsky, "Lower Colorado River," 22.

24. H. T. Corey, "Irrigation and River Control in the Colorado River Delta," 1249–50.

25. Frederick Kershner Jr., "George Chaffey and the Irrigation Frontier," 120–21; Schonfeld, "Early Development of Imperial Valley," 288.

26. Kershner, "George Chaffey and the Irrigation Frontier," 121.

27. Corey, "Irrigation and River Control," 1258, 1269.

28. Schonfeld, "Early Development of Imperial Valley," 290.

29. Department of the Interior, U.S. Geological Survey, *Fourth Annual Report of the Reclamation Service, 1904–5,* 103; Corey, "Irrigation and River Control," 1269.

30. E. C. La Rue, *Colorado River and Its Utilization,* 143; Department of the Interior, U.S. Geological Survey, *Fourth Annual Report of the Reclamation Service,* 102; Corey, "Irrigation and River Control," 1252.

31. Corey, "Irrigation and River Control," 1246.

32. Grunsky, "Lower Colorado River," 23.

33. Department of the Interior, Bureau of Land Management, Tucson District Land Office Tract Books.

34. Richard Wells Bradfute, *The Court of Private Land Claims: The Adjudication of Spanish and Mexican Land Grant Titles, 1891–1904,* 161.

35. *United States v. Coe,* 746.

36. Department of the Interior, Bureau of Land Management, Tucson District Land Office Tract Books.

37. *United States v. Coe,* 747, 745, 753.

38. J. Garnett Holmes, "Soil Survey of the Yuma Area, Arizona," 778.

39. O. P. Bondesson to F. L. Sellew, July 9, 1908, File 11/11, Box 17, Case 74: Tucson Equity Case, National Archives, Pacific Southwest Region, RG 21: Records of the District Courts of the United States for the District of Arizona.

40. Holmes, "Soil Survey," 778.

41. Ibid.

42. Bondesson to Sellew, July 9, 1908, File 11/11, Box 17, Case 74: Tucson Equity Case, National Archives, Pacific Southwest Region, RG 21.

43. According to Worster, the suspended sediment load floating past the site of Yuma amounted to 160 million tons a year (*Rivers of Empire,* 195).

44. Bondesson to Sellew, July 9, 1908, File 11/11, Box 17, Case 74: Tucson Equity Case, National Archives, Pacific Southwest Region, RG 21.

45. Ibid.

46. Holmes, "Soil Survey," 779.

4. THE YUMA PROJECT

1. Worster, *Rivers of Empire,* 209; L. M. Holt, "The Reclamation Service and the Imperial Valley," 72–73.

2. Department of the Interior, U.S. Geological Survey, *First Annual Report of the Reclamation Service,* 63; Department of the Interior, U.S. Geological Survey, *Third Annual Report of the Reclamation Service, 1903–4,* 48.

3. Schonfeld, "Early Development of Imperial Valley," 280, 293.

4. Donald Pisani, *From the Family Farm to Agribusiness: The Irrigation Crusade in California and the West, 1950–1931,* 309.

5. J. B. Lippincott to Senator Thomas Bard, August 8, 1902, Box 9B, File: Irrigation IV, Thomas R. Bard Collection, Huntington Library, San Marino, Calif.

6. Holmes, "Soil Survey," 791; Holmes et al., *Soil Survey, Arizona-California,* 23; J. B. Lippincott, "The Reclamation Service in California," 169.

7. W. H. Code to the secretary of the interior, October 17, 1902, File 154-A: Indian Lands, Box 1088, National Archives, Washington, D.C., RG 115: Records of the Bureau of Reclamation, General Administrative and Project Records, 1902–1919, Yuma Project.

8. W. A. Jones to the secretary of the interior, November 29, 1902, October 27, 1903, ibid.

9. Holmes et al., *Soil Survey, Arizona-California,* 21.

10. A. P. Davis et al. to F. H. Newell, April 8, 1904, File 11/11, Box 17, Case 74: Tucson Equity Case, National Archives, Pacific Southwest Region, RG 21.

11. Corey, "Irrigation and River Control," 1218.

12. Grunsky, "Lower Colorado River," 50.

13. Schonfeld, "Early Development of Imperial Valley," 295.

14. Quoted in ibid.

15. Ibid.

16. Ibid., 296–97.

17. Lippincott to Bard, March 16, 1904, Box 9B: File: Irrigation IV, Bard Collection; bill quoted in Corey, "Irrigation and River Control," 1274.

18. Lippincott to Bard, March 16, 1904, Box 9B: File: Irrigation IV, Bard Collection.

19. Department of the Interior, U.S. Geological Survey, *Third Annual Report of the Reclamation Service,* 48–49.

20. Lippincott to Governor George Pardee, March 17, 1904, Box 79, Folder: Lippincott, J. B., George Cooper Pardee Correspondence and Papers, 1890–1941, Bancroft Library, University of California–Berkeley.

21. Schonfeld, "Early Development of Imperial Valley," 291.

22. *Address of Hon. A. H. Heber to the Settlers of Imperial Valley, July 25, 1904,* 4.

23. Corey, "Irrigation and River Control," 1274.

24. Quoted in ibid.

25. *Address of Hon. A. H. Heber,* 4–5.

26. *U.S. Statutes at Large* 33 (1904): 224.

27. Lippincott to Newell, April 10, 1904, File 546: Yuma Valley WUA to December 31, 1909, Box 1100, National Archives, Washington, D.C., RG 115.

28. Norris Hundley Jr., *Dividing the Waters: A Century of Controversy Between the United States and Mexico,* 40.

29. M. Winsor to the secretary of the interior, May 3, 1904, File 11/11, Box 17, Case 74: Tucson Equity Case, National Archives, Pacific Southwest Region, RG 21.

30. Ibid.; Lippincott to Newell, April 10, 1904, File 546: Yuma Valley WUA to December 31, 1909, Box 1100, National Archives, Washington, D.C., RG 115.

31. Elwood Mead and Burton P. Fleming, "Report of the Local Board of Review, U.S. Reclamation Service, Yuma Project, Arizona-California," 10.

32. Ibid., 11.

33. Bondesson to Sellew, July 9, 1908, File 11/11, Box 17, Case 74: Tucson Equity Case, National Archives, Pacific Southwest Region, RG 21.

34. J. C. Avakian to Homer Hamlin, November 14, 1904, File 154-A: Indian Lands, Box 1088, National Archives, Washington, D.C., RG 115.

35. Corey, "Irrigation and River Control," 1272.

36. Department of the Interior, U.S. Geological Survey, *Fourth Annual Report of the Reclamation Service,* 102.

37. *Address of Hon. A. H. Heber,* 4; Department of the Interior, U.S. Geological Survey, *Fourth Annual Report of the Reclamation Service,* 102–3.

38. *Address of Hon. A. H. Heber,* 13.

39. Department of the Interior, U.S. Geological Survey, *Fourth Annual Report of the Reclamation Service,* 107.

40. Ibid., 103.

41. Corey, "Irrigation and River Control," 1271.

42. Ibid., 1250.

43. Department of the Interior, U.S. Geological Survey, *Fourth Annual Report of the Reclamation Service,* 103.

44. *The Unfriendly Attitude of the United States Government Towards the Yuma Valley, Arizona,* 28.

45. Corey, "Irrigation and River Control," 1273 (including quote).

46. Holt, "Reclamation Service," 73.

47. *Address of Hon. A. H. Heber,* 13.

48. Corey, "Irrigation and River Control," 1289–90; Department of the Interior, U.S. Geological Survey, *Fourth Annual Report of the Reclamation Service,* 107.

49. Corey, "Irrigation and River Control," 1290.

50. Grunsky, "Lower Colorado River," 16.

51. Corey, "Irrigation and River Control," 1292–93; Grunsky, "Lower Colorado River," 30.

52. Corey, "Irrigation and River Control," 1313; Grunsky, "Lower Colorado River," 45.

53. Corey, "Irrigation and River Control," 1296.

54. Grunsky, "Lower Colorado River," 3.

55. Ibid., 4.

56. Corey, "Irrigation and River Control," 1362–63.

5. ALLOTMENT

1. Corey, "Irrigation and River Control," 1238.

2. Grunsky, "Lower Colorado River," 36; Mead and Fleming, "Report of the Local Board of Review," 14–15.

3. Grunsky, "Lower Colorado River," 48.

4. Corey, "Irrigation and River Control," 1244.

5. Lippincott to Newell, June 27, 1904, File 154-A: Indian Lands, Box 1088, National Archives, Washington, D.C., RG 115.

6. Newell to Davis, July, 3, 1905; acting director of the U.S. Geological Survey to the secretary of the interior, July 15, 1905, ibid.

7. Census of the Yuma Indians of Fort Yuma School, California Agency, taken by John S. Spear, superintendent, and U.S. Indian agent, June 30, 1905, ibid.

8. Director, U.S. Geological Survey, to the secretary of the interior, February 28, 1906, ibid.

9. Davis to Lippincott, February 23, 1906, ibid.

10. Lippincott to Newell, June 27, 1904, July 25, 1905, ibid.

11. Acting commissioner of Indian affairs to the secretary of the interior, October 30, 1905, ibid.

12. Spear to the commissioner of Indian affairs, September 14, 1905, ibid.

13. Director, Yuma County Water Users Association, to the secretary of the interior, December 20, 1904, File 546: Yuma Valley WUA to December 31, 1909, Box 1100, ibid.

14. Hamlin to Lippincott, November 29, 1905, December 6, 1905, File 154-A: Indian Lands, Box 1088, ibid.

15. Hamlin to Lippincott, December 6, 1905, ibid.

16. Lippincott to Newell, January 12, 1906, ibid.

17. Frederick E. Hoxie, *A Final Promise: The Campaign to Assimilate the Indians, 1880–1920*, 152.

18. Quoted in ibid., 158.

19. Pisani, *Water and the American Government*, 200.

20. Hoxie, *Final Promise*, 155. In *Lone Wolf v. Hitchcock* (187 U.S. [1903]: 565–66), the Court recognized an almost absolute congressional power over Indian affairs that was virtually exempt from judicial oversight, arguing that Congress had plenary power over Indian property "by reason of its exercise of guardianship over their interest."

21. Levi Chubbuck to the secretary of the interior, April 6, 1907, File 154-A: Indian Lands, Box 1088, National Archives, Washington, D.C., RG 115.

22. Ira C. Deaver to the commissioner of Indian affairs, April 11, 1907, ibid.

23. Code to the secretary of the interior, June 12, 1907, ibid.

24. Bee mistakenly contends that much of the "good farmland" in what would become known as the Bard district in the eastern part of the reservation had already been sold to Anglos before the Quechans received their allotments, and that many of the Quechans ended up with "worthless semi-desert tracts" in the western part of the reservation (*Crosscurrents Along the Colorado*, 66). However, the location of Quechan allotments was determined long before non-Indians were allowed to purchase land in the Bard district, and it will be shown in chapter 8 that many of the Bard tracts were much less suited to irrigated farming than the tracts that were allotted the Quechans, which actually was the best of the reservation land.

25. Douglas D. Graham to the commissioner of Indian affairs, August 29, 1909, File 154-A: Indian Lands, Box 1088, National Archives, Washington, D.C., RG 115.

26. McCool, *Command of the Waters*, 142.

27. S. M. Brosius to the commissioner of Indian affairs, August 28, 1909, File 154-A: Indian Lands, Box 1088, National Archives, Washington, D.C., RG 115.

28. Brosius, "A Plea for Justice for the Yuma Indians," to *California Independent*, August 26, 1909, ibid.

29. Brosius to the commissioner of Indian affairs, August 28, 1909, ibid.

30. Brosius, "Plea for Justice."

31. J. M. Ocheltree to *California Independent*, August 26, 1909; Joseph E. Johnson to the commissioner of Indian affairs, September 15, 1909; "Want Just Allotments to Indians," *Los Angeles Express*, October 6, 1909, all in File 154-A: Indian Lands, Box 1088, National Archives, Washington, D.C., RG 115.

32. Capt. Simon Miguel to the commissioner of Indian affairs, September 17, 1909, ibid.

33. Ibid.

34. Charles Curtis to the commissioner of Indian affairs, September 15, 1909, ibid.

35. Newell to the commissioner of Indian affairs, October 14, 1909, ibid.

36. Ibid.

37. Acting commissioner of Indian affairs to Newell, October 20, 1909, ibid.

38. Ibid., October 21, 1909.

39. Newell to the acting commissioner of Indian affairs, October 22, 1909, ibid.

40. Louis C. Hill to Newell, December 27, 1909, ibid.

41. Newell to Hill, January 5, 1910; Hill to Newell, January 20, 1910; Newell to the commissioner of Indian affairs, February 17, 1910, ibid.

42. U.S. Congress, Senate, "Allotment of Irrigable Lands to Indians on Colorado River and Yuma Reservations"; *U.S. Statutes at Large* 36 (1911): 1058–63.

43. Charles E. Roblin to the commissioner of Indian affairs, July 21, 1912, File 154-A: Indian Lands, 1910–1919, Box 1088, National Archives, Washington, D.C., RG 115.

44. Bee, *Crosscurrents Along the Colorado,* 50.

45. Hoxie, *Final Promise,* 162–63.

6. BARD

1. Newell to Hill, January 14, 1909, File 560-C: Townsites, Box 1105, National Archives, Washington, D.C., RG 115.

2. Hill to Newell, October 3, 1907; R. H. Forbes to Davis, October 1, 1907, File 560-B: Classification of Lands, Box 1104, National Archives, Washington, D.C., RG 115.

3. Davis to Forbes, October 7, 1907, ibid.

4. W. A. Peterson to C. S. Scofield, June 1, 1908; settlers on public lands under the Yuma Project to the secretary of the interior, June 13, 1908, ibid.

5. Settlers on public lands under the Yuma Project to the secretary of the interior, June 13, 1908, ibid.

6. Ibid.

7. Sellew to Newell, June 2, 1908, ibid.

8. Ibid.

9. Ibid.

10. Forbes to Newell, June 3, 1908, ibid.

11. Newell to Sellew, June 11, 1908, ibid.

12. Ibid.; Sellew to Newell, June 20, 1908, ibid.

13. Chief engineer, U.S. Reclamation Service, to Sellew, September 21, 1908; Sellew to Newell, September 9, 1908, File 560: Classification of Lands, Box 1100, ibid.; "Official Farm Unit Plat—Indian Reservation—Yuma Project," File 560-A1, Box 1101, ibid.

14. "To the Yuma Landseeker," n.d., File 154-A: Indian Lands, 1910–1919, Box 1088, National Archives, Washington, D.C., RG 115.

15. Newell to the secretary of the interior, February 27, 1909; "Line Hitters Standing Pat: Yuma Land Seekers Still in Aggressive Mood; No Change in Department Plan for Filings; Indications Are for Fierce Rush at Opening," *Los Angeles Times,* February 22, 1910, ibid.

16. Acting commissioner, Government Land Office, to register and receiver, Los Angeles, February 5, 1910, File 560-A: Public Notices Thru 1915, Box 1101, ibid.

17. Sellew to Newell, March 26, 1910, File 154-A: Indian Lands, 1910–1919, Box 1088, ibid.

18. "Farm Only for One in Ten," *Los Angeles Times*, March 23, 1910, File 153-11: Newspaper Clippings, Box 1087, ibid.

19. "Lucky Winners of Land," File 154-A: Indian Lands, 1910–1919, Box 1088, ibid.

20. W. O. Harris to the secretary of the interior, April 5, 1918, File 154: Lands General, Box 1088, National Archives, Washington, D.C., RG 115.

21. Donald Pisani, "Reclamation and Social Engineering in the Progressive Era," 54.

22. Newell to the supervising engineer, June 2, 1911, File 560-C: Townsites, Box 1105, National Archives, Washington, D.C., RG 115.

23. Pisani, "Reclamation and Social Engineering," 56.

24. L. M. Lawson to the chief of construction, April 19, 1917, File 560-C: Townsites, Box 1105, National Archives, Washington, D.C. RG 115.

7. YUMA VALLEY TRAVAILS

1. Mead and Fleming, "Report of the Local Board of Review," 13.

2. Board of Engineers to Newell, May 4, 1905, File 150-A: Acquisitions of Land, Box 1078, National Archives, Washington, D.C., RG 115.

3. Lippincott to Newell, September 29, December 13, 1904, File 546: Yuma Valley WUA to December 31, 1909, Box 1100, ibid.

4. Board of Engineers to Newell, May 4, 1905, File 150-A: Acquisitions of Land, Box 1078, ibid.

5. Mead and Fleming, "Report of the Local Board of Review," 14.

6. Irrigation Land and Improvement Co. to the secretary of the interior, March 11, 1905, File 150-A: Acquisitions of Land, Box 1078, National Archives, Washington, D.C., RG 115.

7. Board of Engineers to Newell, May 4, 1905, ibid.

8. Department of the Interior, *Fifth Annual Report of the Reclamation Service, 1906*, 30.

9. *Unfriendly Attitude*, 32.

10. Bondesson to Sellew, July 9, 1908, File 11/11, Box 17, Case 74: Tucson Equity Case, National Archives, Pacific Southwest Region, RG 21.

11. Petitioners to the secretary of the interior, March 19, 1906, File 150: Purchase of Lands and Right of Way General Thru 1908, Box 1078, National Archives, Washington, D.C., RG 115.

12. Ibid.

13. Lippincott to Newell, April 9, 1906, ibid.

14. *Unfriendly Attitude*, 18.

15. Lippincott to Newell, April 9, 1906, File 150: Purchase of Lands and Right of Way General Thru 1908, Box 1078, National Archives, Washington, D.C., RG 115.

16. *Unfriendly Attitude,* 18.

17. Charles D. Walcott to the secretary of the interior, May 15, 1906, File 150: Purchase of Lands and Right of Way General Thru 1908, Box 1078, National Archives, Washington, D.C., RG 115.

18. Secretary of the interior to George Turner, May 29, 1906, ibid.

19. Ibid.

20. Bondesson to Sellew, July 9, 1908, File 11/11, Box 17, Case 74: Tucson Equity Case, National Archives, Pacific Southwest Region, RG 21.

21. *Unfriendly Attitude,* 23.

22. Ibid., 14.

23. Ibid., 15.

24. Ibid., 16.

25. Consolidated Water Users Association to Theodore Roosevelt, January 26, 1907, File 150: Purchase of Lands and Right of Way General Thru 1908, Box 1078, National Archives, Washington, D.C., RG 115.

26. Ibid.

27. *Unfriendly Attitude,* 49–51.

28. Bondesson to Sellew, July 9, 1908, File 11/11, Box 17, Case 74: Tucson Equity Case, National Archives, Pacific Southwest Region, RG 21.

29. R. S. Fessenden, "Operation and Maintenance Report, 1916, Yuma Project, Arizona-California," 4.

30. Sellew, U.S. Reclamation Service, "Yuma Project Historical Sketch, 1902–1912," 94.

31. U.S. Reclamation Service, Yuma Project, Arizona-California, "Data Prepared for Board of Engineers, U.S.A., September 1910," 53.

32. Department of the Interior, *Tenth Annual Report of the Reclamation Service, 1910–1911,* 76.

33. J. S. Garvin et al. to the secretary of the interior, n.d., File 546: Yuma Valley WUA to December 31, 1909, Box 1100, National Archives, Washington, D.C., RG 115.

34. Ibid.

35. Department of the Interior, *Thirteenth Annual Report of the Reclamation Service, 1913–1914,* 16.

36. Reisner, *Cadillac Desert,* 120.

37. Mead and Fleming, "Report of the Local Board of Review," 17.

38. "Water Users Draft a Strong Petition," *Yuma Examiner,* April 10, 1911, File 546: Yuma Valley WUA to December 31, 1909, Box 1100, National Archives, Washington, D.C., RG 115.

39. Department of the Interior, *Eleventh Annual Report of the Reclamation Service, 1911–1912,* 36.

8. DISTRESS AND DISCONTENT

1. Quoted in Department of the Interior, U.S. Geological Survey, *First Annual Report of the Reclamation Service,* 44.

2. Sellew to Newell, May 2, 1910, File 154: Lands General, Box 1088, National Archives, Washington, D.C., RG 115; U.S. Reclamation Service, Yuma Project, Arizona-California, "Data Prepared," 23.

3. Newell to the secretary of the interior, n.d., File 560-A1: Land Classification, Payment Under Public Notice Thru March 31, 1915, Box 1101, National Archives, Washington, D.C., RG 115.

4. Compiled from Department of the Interior, Bureau of Land Management, Historical Index, Township 15 South, Range 23 East, S.B.M. and Township 16 South, Range 23 East, S.B.M.

5. Newell, *Irrigation Management,* 3.

6. Homesteaders to the secretary of the interior, May 2, 1911, File 560-A1: Land Classification, Payment Under Public Notice Thru March 31, 1915, Box 1101, National Archives, Washington, D.C., RG 115.

7. Hill to Newell, May 19, June 2, 1909, File 560-A: Public Notices, Thru 1915, Box 1101, National Archives, Washington, D.C., RG 115.

8. Richmond Wisehart to John D. Works, June 6, 1911; Sellew to Hill, June 8, 1911, File 560-A1: Land Classification, Payment Under Public Notice Thru March 31, 1915, ibid.

9. Davis to Hill, November 14, 1911, ibid.

10. Newell to the secretary of the interior, n.d., ibid.; "Yuma Project, California, Department of the Interior, Public Notice," March 8, 1912, File 560-A: Public Notices, ibid.

11. Arizona-California, Yuma Project, "Operation and Maintenance Report for 1913," 3.

12. Morris Bien to the secretary of the interior, February 15, 1913, File 560-A: Public Notices, Box 1101, National Archives, Washington, D.C., RG 115.

13. Mead and Fleming, "Report of the Local Board of Review," 55–56.

14. Fiege, *Irrigated Eden,* 31.

15. O. B. Judd to the secretary of the interior, August 20, 1912, File 154: Lands General, Box 1088, National Archives, Washington, D.C., RG 115.

16. Acting secretary of the interior to Judd, September 14, 1912; Sellew to Newell, September 23, 1912, ibid.; "Yuma Project, California, Department of the Interior, Order," March 6, 1913, File 560-A: Public Notices, Box 1101, ibid.

17. Holmes et al., *Soil Survey, Arizona-California,* 14.

18. Davis et al. to Newell, April 8, 1904; Newell to Yuma County Water Users Association, May 3, 1904, File 11/11, Box 17, Case 74: Tucson Equity Case, National Archives, Pacific Southwest Region, RG 21.

19. "Yuma Irrigation Project, Arizona-California: Information Compiled by United States Reclamation Service, October 1, 1909," 27.

20. Mead and Fleming, "Report of the Local Board of Review," 61.

21. Department of the Interior, *Thirteenth Annual Report of the Reclamation Service,* 21.

22. Quoted in Mead and Fleming, "Report of the Local Board of Review," 27.

23. Ibid.

24. Holmes et al., *Soil Survey, Arizona-California*, 21.

25. Arizona-California, Yuma Project, "Operation and Maintenance Report for 1913," 4.

26. "Yuma Irrigation Project, Arizona-California," 28.

27. Mead and Fleming, "Report of the Local Board of Review," 56; Department of the Interior, *Fifteenth Annual Report of the Reclamation Service, 1915–1916*, 75.

28. Mead and Fleming, "Report of the Local Board of Review," 83.

29. Department of the Interior, *Fourteenth Annual Report of the Reclamation Service*, 8. By the 1920s, reclamation farmers had repaid only a small fraction of what had been spent on building dams and canals, and the idea that federal reclamation could ever become a self-supporting program had long since been abandoned (Pisani, "Federal Reclamation and the American West," 395).

30. Board of Arbitration on Indian Matters to Reclamation Commission and commissioner of Indian affairs, February 5, 1915, File 560-A1: Payments Under Public Notice, Indian Lands, Box 1101, National Archives, Washington, D.C., RG 115.

31. R. E. Blair, *The Work of the Yuma Reclamation Project Experiment Farm in 1918*, 7.

32. Department of the Interior, *Fifteenth Annual Report of the Reclamation Service*, 78.

33. Ibid.; U.S. Reclamation Service, Yuma Project, "Historical Sketch: Appendix No. 2 Covering the Period From January 1 to December 31, 1914," 18.

34. Mead and Fleming, "Report of the Local Board of Review," 60, 105–10.

35. Ibid., 70.

36. Mead and Fleming, "Proceedings, Local Board of Review, Yuma Project, Arizona-California: Part I, California Lands," 4–5.

37. Mead and Fleming, "Report of the Local Board of Review," 108.

38. S. P. Huss and P. T. Taylor to Lawson, April 8, 1916, File 560-A6: Payment Under Public Notice—Postponement of Charges on Account of Seepage, Box 1104, National Archives, Washington, D.C., RG 115.

39. Lawson to Newell, June 28, 1916, ibid.

40. Huss to Lawson, July 11, 1916, ibid.

41. Lawson to Newell, August 5, 1916, ibid.

42. W. W. Schlecht to the chief of construction, March 26, 1918, ibid.

43. Ibid.

44. Holmes, "Soil Survey," 791.

45. William E. Smythe, "An International Wedding," 288–89.

46. Department of the Interior, *Thirteenth Annual Report of the Reclamation Service*, 76–77.

47. Fessenden, "Operation and Maintenance Report, 1916," 9.

48. Department of the Interior, *Sixteenth Annual Report of the Reclamation Service, 1916–1917*, 62.

49. Harris to the secretary of the interior, March 14, 1918, File 154: Lands General, Box 1088, National Archives, Washington, D.C., RG 115.

50. F. O. Youngs et al., *Soil Survey of the Yuma-Wellton Area, Arizona-California*, 11.

51. Harris to the secretary of the interior, April 5, 1918, File 154: Lands General, Box 1088, National Archives, Washington, D.C., RG 115.

52. "Annual Report, 1912," 1, Box 44: Superintendents' Annual Reports, 1911–1920, National Archives, Pacific Southwest Region, RG 75: Records of the Bureau of Indian Affairs, Fort Yuma Agency.

53. Ibid., 2. Even though Indian trust patents were not issued until February 1914, the Quechans were allowed to take up their allotments after the allotment process was completed in April 1912.

54. "Annual Report, 1914," 17, ibid.

55. "Annual Report, 1911," 2, ibid.

56. Leonard A. Carlson, *Indians, Bureaucrats, and Land: The Dawes Act and the Decline of Indian Farming*, 91.

57. Memorandum for the commissioner of Indian affairs, "On the Irrigation of the Indian Portion of the Yuma Project," August 6, 1914, File 154-A: Indian Lands, 1910 Thru 1919, Box 1088, National Archives, Washington, D.C., RG 115.

58. Memorandum, Department of the Interior, United States Reclamation Service, March 21, 1913, ibid.

59. Memorandum for the commissioner of Indian affairs, "On the Irrigation of the Indian Portion of the Yuma Project," August 6, 1914, ibid.

60. Board of Arbitration on Indian Matters to Reclamation Commission and commissioner of Indian affairs, February 5, 1915, File 560-A1: Payments Under Public Notice, Indian Lands, Box 1101, ibid.

61. Ibid.

62. Memorandum for the commissioner of Indian affairs, "On the Irrigation of the Indian Portion of the Yuma Project," August 6, 1914, File 154-A: Indian Lands, 1910 Thru 1919, Box 1088, ibid.

63. "Annual Report, 1911," 2; "Annual Report, 1912," 1, both in Box 44: Superintendents' Annual Reports, 1911–1920, National Archives, Pacific Southwest Region, RG 75.

64. Quoted in Bee, *Crosscurrents Along the Colorado*, 69.

65. Ibid.

66. Leibhardt, "Allotment Policy," 95.

67. Hoxie, *Final Promise*, 79.

68. "Annual Report, 1915," 1; "Annual Report, 1920," 25, both in Box 44: Superintendents' Annual Reports, 1911–1920, National Archives, Pacific Southwest Region, RG 75; Irrigation Data, Long Range Program, Yuma, 4, Irrigation and Water Rights Case Files, ibid.

69. Mead and Fleming, "Report of the Local Board of Review," 83.

70. Hoxie, *Final Promise*, 184.

71. "Annual Report, 1922," 2, Box 44: Superintendents' Annual Reports, 1921–1924, RG 75.

72. Bee, *Crosscurrents Along the Colorado*, 70.

73. "Annual Report, 1915," 19, Box 44: Superintendents' Annual Reports, 1911–1920, National Archives, Pacific Southwest Region, RG 75.

74. "Annual Reports, 1916–1920," ibid.

75. Porter J. Preston and Charles A. Engle, "Report of Advisors on Irrigation on Indian Reservations," 31.

76. "Annual Report, 1917," 19, Box 44: Superintendents' Annual Reports, 1911–1920, National Archives, Pacific Southwest Region, RG 75; "Annual Report, 1923," 14–15, Box 44: Superintendents' Annual Reports, 1921–1924, National Archives, Pacific Southwest Region, RG 75.

77. "Annual Report, 1923," 13, Box 44: Superintendents' Annual Reports, 1921–1924, National Archives, Pacific Southwest Region, RG 75.

78. Bee, *Crosscurrents Along the Colorado,* 72.

79. "Annual Report, 1920," 8, Box 44: Superintendents' Annual Reports, 1911–1920, National Archives, Pacific Southwest Region, RG 75.

80. Bee, *Crosscurrents Along the Colorado,* 72.

81. Ibid., 70.

82. Ibid., 86.

83. Hoxie, *Final Promise,* 186, 187.

84. Arizona-California, Yuma Project, "Operation and Maintenance Report for 1913," 4.

85. Hill to Newell, June 10, 1913, File 560-A: Public Notices Thru 1915, Box 1101, National Archives, Washington, D.C., RG 115.

86. Brit Allan Storey, *The Bureau of Reclamation's Yuma Valley Railroad,* 3–4.

87. Ibid., 13.

88. Ibid., 23.

89. Ibid., 53.

90. "Yuma Project Annual Project History and O. and M. Report, 1921," 58.

91. Storey, *Yuma Valley Railroad,* 48.

92. Robinson, *Water for the West,* 40.

93. Lampen, *Economic and Social Aspects,* 67.

94. Mead and Fleming, "Report of the Local Board of Review," 110–11.

95. Ibid., 112.

96. Ibid., 114.

97. Ibid., 116.

98. Department of the Interior, *Eighteenth Annual Report of the Reclamation Service, 1918–1919,* 20.

99. File 4/11, Box 17, Case 74: Tucson Equity Case, National Archives, Pacific Southwest Region, RG 21.

100. A common complaint among Reclamation project farmers was that the Reclamation Service intentionally obscured the per-acre cost of reclamation by predicting that far more land could be irrigated within each project than was realistic (Pisani, *Water and the American Government,* 116).

101. Department of the Interior, *Fifteenth Annual Report of the Reclamation Service,* 7.

102. Testimony of Arthur P. Davis, File 3/11, Box 17, Case 74: Tucson Equity Case, National Archives, Pacific Southwest Region, RG 21.

103. Department of the Interior, *Eighteenth Annual Report of the Reclamation Service*, 25.

104. "Yuma Project Annual Project History and O. and M. Report, 1920," 51.

105. Department of the Interior, *Eighteenth Annual Report of the Reclamation Service*, 13.

9. RECLAMATION AND RETROSPECT

1. *Yuma Premier Project of the U.S.R.S.*, 6.

2. Mead and Fleming, "Report of the Local Board of Review," 118.

3. "Yuma Project Annual Project History and O. and M. Report, 1925," 98.

4. *Yuma, Arizona, U.S.R.S. Premier Project*, 61.

5. Department of the Interior, *Thirteenth Annual Report of the Reclamation Service*, 77, 440.

6. "Yuma Irrigation Project, Arizona-California," 19.

7. Department of the Interior, *Fourteenth Annual Report of the Reclamation Service*, 17.

8. Ibid., 60.

9. Department of the Interior, *Nineteenth Annual Report of the Reclamation Service, 1919–1920*, 91.

10. *Yuma Premier Project of the U.S.R.S.*, 15.

11. "Yuma Project Annual Project History and O. and M. Report, 1919," 2; "Yuma Project Annual Project History and O. and M. Report, 1925," 97.

12. Pisani, *Water and the American Government*, 123.

13. Department of the Interior, *Sixteenth Annual Report of the Reclamation Service*, 63; Department of the Interior, *Eighteenth Annual Report of the Reclamation Service*, 21.

14. "Yuma Project Annual Project History and O. and M. Report, 1919," 2.

15. "Yuma Project Annual History, 1917," 48.

16. Department of the Interior, *Eighteenth Annual Report of the Reclamation Service*, 21.

17. Department of the Interior, Bureau of Reclamation, *Twenty-third Annual Report*, 46.

18. Department of the Interior, Bureau of Reclamation, *Yuma Federal Reclamation Project, Arizona-California*, 10.

19. Pisani, *Water and the American Government*, 129.

20. "Yuma Project Annual Project History and O. and M. Report, 1922," 41; "Yuma Project Annual Project History and O. and M. Report, 1924," 11.

21. Department of the Interior, Bureau of Reclamation, *Twenty-third Annual Report*, 46.

22. U.S. Congress, Senate, *Federal Reclamation by Irrigation*, 229, 96.

23. "Yuma Project Annual Project History and O. and M. Report, 1924," 11.

24. Lewis Meriam, *The Problem of Indian Administration*, 7.

25. Department of the Interior, *Sixteenth Annual Report of the Reclamation Service*, 62.

26. "Yuma Project O. and M. Report, 1914," 12.

27. Department of the Interior, *Nineteenth Annual Report of the Reclamation Service*, 90.

28. "Yuma Project Annual Project History and O. and M. Report, 1919," 28.

29. Department of the Interior, *Thirteenth Annual Report of the Reclamation Service*, 76; Department of the Interior, Bureau of Reclamation, *Twenty-third Annual Report of the Reclamation Service*, 46.

30. Lampen, *Economic and Social Aspects*, 105–6.

31. William E. Warne, "Land Speculation," 176.

32. Carlson, *Indians, Bureaucrats, and Land*, 79; Lewis, *Neither Wolf nor Dog*, 16.

33. Department of the Interior, Office of Indian Affairs, *Indian Land Tenure, Economic Status, and Population Trends*, 23.

34. "Yuma Project Annual Project History and O. and M. Report, 1925," 96.

35. "Irrigation Data Long-Range Program, Yuma," 4.

36. Aschmann, "Head of the Colorado Delta," 258.

37. "Irrigation Data Long-Range Program, Yuma," 5.

38. Department of the Interior, Bureau of Indian Affairs, Land Titles and Records Office, Albuquerque.

39. Although today the BIA, which arranges the allotment leases, attempts to achieve unanimous agreement for each lease, lacking that, majority ownership rules regarding the lease agreements (telephone interview, Brian Golding Sr., economic development director, Quechan Indian Tribe, Fort Yuma Indian Reservation, September 10, 2007).

40. F. H. Newell, "The Reclamation of the West," 828.

41. U.S. Congress, Senate, *Federal Reclamation by Irrigation*, xii.

42. R. P. Teele, *The Economics of Land Reclamation in the United States*, 112–13.

43. F. H. Newell, "National Efforts at Home Making," 531.

44. U.S. Congress, Senate, *Federal Reclamation by Irrigation*.

45. Meriam, *Problem of Indian Administration*.

46. The committee advised that the agricultural and economic side of federal reclamation should be given more attention, and that the Reclamation Service should be completely reorganized and overhauled. The service had already partially been restructured in the summer of 1923, when it became the Bureau of Reclamation; then, in the spring of 1924, the bureau was divided into three main departments—Engineering, Finance, and Farm Economics. In May 1924, Elwood Mead became the commissioner of the reorganized bureau.

47. U.S. Congress, Senate, *Federal Reclamation by Irrigation*, xii, 35.

48. Ibid., xiii.

49. Imre Sutton, *Indian Land Tenure: Bibliographical Essays and a Guide to the Literature*, 68.

50. Meriam, *Problem of Indian Administration,* 449.

51. Preston and Engle, "Report of Advisors," 33 (quote), 34.

52. Quoted in Department of the Interior, Office of Indian Affairs, *Indian Land Tenure,* 7–8 (quote), 6, 11.

53. Preston and Engle, "Report of Advisors," 21.

54. Lewis, *Neither Wolf nor Dog,* 7.

55. Pisani, "Reclamation and Social Engineering."

56. Telephone interview with Golding, September 10, 2007.

57. Nelson D. Schwartz, "Far From the Reservation, but Still Sacred?" *New York Times,* August 12, 2007, BU1. A much larger casino, to be located just off Interstate 8 in California, was approved by the California legislature in 2006; its planned opening is February 2009, and it is expected to employ between six and eight hundred people (telephone interview with Golding, September 10, 2007).

58. Dan Morain, "Casino Tribes Try to Keep Entire Pot," *Los Angeles Times,* August 22, 2006, A1. The primary sources of income for tribal members include employment or public assistance or both, followed by relatively small per capita distributions of gaming revenues, and allotment leases. In lieu of larger per capita gaming distributions, the tribe's priority is to invest in social services programs and infrastructure to meet the tribe's current and future needs (telephone interview with Golding, September 10, 2007).

59. Personal interview, Charles Sanchez, resident director, Yuma Agricultural Center, December 20, 1999.

60. Lewis, *Neither Wolf nor Dog,* 168.

61. Ibid., 20, 169.

62. Ibid., 69 (quote).

10. EPILOGUE

1. Worster, "Thinking Like a River," 59.

2. Philip L. Fradkin, *A River No More: The Colorado River and the West.*

3. Robinson, *Water for the West,* 49.

4. Ibid., 50.

5. Worster, *Rivers of Empire,* 208.

6. Robert Jerome Glennon and Peter W. Culp, "The Last Green Lagoon: How and Why the Bush Administration Should Save the Colorado River Delta," 918.

7. Fradkin, *A River No More,* 187.

8. Ibid.

9. Ibid., 189.

10. Glennon and Culp, "Last Green Lagoon," 919–20.

11. Fradkin, *A River No More,* 299.

12. Evan R. Ward, *Border Oasis: Water and the Political Ecology of the Colorado River Delta, 1940–1975,* 23.

13. Fradkin, *A River No More,* 300, 301.

14. Glennon and Culp, "Last Green Lagoon," 913–15 (quote on p. 915).

15. Ward, *Border Oasis,* 36.

16. Ibid., 51.

17. Fradkin, *A River No More,* 302.

18. Worster, *Rivers of Empire,* 321.

19. Fradkin, *A River No More,* 303.

20. Ward, *Border Oasis,* 73 (quote), 87.

21. Ibid., 70.

22. Fradkin, *A River No More,* 302.

23. International Boundary Water Commission decisions are referred to as Minutes (Glennon and Culp, "Last Green Lagoon," 915).

24. Ward, *Border Oasis,* 116.

25. Worster, *Rivers of Empire,* 322.

26. Joe Gelt, "Basin States Consider Ways to Share Colorado River Shortages," 8.

27. "Yuma Desalter Attracts International Interest."

28. Randal C. Archibold and Kirk Johnson, "The Arid West No Longer Waits for Rain," *New York Times,* April 4, 2007, A1.

29. Robert Glennon and Jennifer Pitt, "Our Water Future Needs Creativity," *Arizona Republic,* May 10, 2004, B7.

30. Ibid.; Glennon and Culp, "Last Green Lagoon," 908.

31. For a discussion of various proposals to save the delta, see Glennon and Culp, "Last Green Lagoon," 951–92. In 2005, a committee composed of individuals who were on opposing sides of the issue over restarting the Yuma desalter drew up a series of recommendations offering potential solutions to the dispute. See Joe Gelt, "Opposing Sides Find Common Ground in Yuma Desalter Controversy." See also "Balancing Water Needs on the Lower Colorado River: Recommendations of the Yuma Desalting Plant/Cienega de Santa Clara Workgroup, April 22, 2005."

32. Glennon and Culp, "Last Green Lagoon," 937.

33. Philip R. Pryde, "The Southern California Water Transfers and the Salton Sea," 1.

34. Fradkin, *A River No More,* 267; Tony Perry, "Southland Share of Water to Be Cut As Deal Collapses," *Los Angeles Times,* January 1, 2003, A1; Tony Perry, "U.S. Raises Stakes in Water Battle," *Los Angeles Times,* December 28, 2002, B1.

35. Seth Hettena, "Imperial Valley Is Thirsting for a Water Fight," *Los Angeles Times,* March 2, 2003; Tony Perry, "Imperial Farmers Should Get Less Water," *Los Angeles Times,* July 4, 2003, B1.

36. Pryde, "Southern California Water Transfers," 6. IID farmers will reduce their water use by changing their water practices, installing updated equipment, and taking other steps to reduce the amount of irrigation water they require. Additional water will be saved by improving the efficiency of the IID's distribution canals. These conservation measures will be paid for by the SDCWA's member districts that will buy the conserved water. San Diego and the IID also plan to replace sections of the present All-American Canal with a new one, excavated alongside the existing one and lined with concrete to make it impervious to seepage. The water that currently leaks out of the unlined canal, twenty-two billion gallons a year, percolates underground,

migrates south across the border, and reemerges in the Mexicali Valley. This windfall of water has helped sustain farmers in the Mexicali Valley for decades; however, it will ultimately be transferred to San Diego. According to Matt Jenkins, "A profound paradox stands at the heart of the logic of efficiency: Increased efficiency creates losers as well as winners, and the victims often inhabit places far beyond the public eye" ("The Efficiency Paradox," 9).

37. Perry, "U.S. Raises Stakes."

38. Ibid.

39. Tony Perry, "Better Water Deal Is Sought," *Los Angeles Times,* December 30, 2002, B1.

40. Pryde, "Southern California Water Transfers," 8.

41. Tony Perry, "Officials Sign Deal to End Feud, Divide Up Water," *Los Angeles Times,* October 17, 2003, B1.

42. Kirk Johnson and Dean E. Murphy. "Drought Settles in, Lake Shrinks, and West's Worries Grow," *New York Times,* May 2, 2004, A1.

43. Bettina Boxall, "By Every Measure, It's Been Dry," *Los Angeles Times,* March 31, 2007, A1; Glen MacDonald, "Hot and Dry—for Decades," *Los Angeles Times,* July 13, 2007, A21; Glen MacDonald, Sigrid Rian, and Hugo Hildalgo, "Southern California and the 'Perfect Drought.'"

44. Tony Davis, "Pessimism Grows on CAP Water Shortage," *Arizona Daily Star,* February 26, 2007, A7.

45. National Research Council, Committee on Scientific Bases of Colorado River Basin Water Management, *Colorado River Basin Water Management: Evaluating and Adjusting to Hydroclimatic Variability,* 108, 3, 88.

46. MacDonald, "Hot and Dry—for Decades."

47. Connie A. Woodhouse, Stephen T. Gray, and David M. Meko, "Updated Streamflow Reconstructions for the Upper Colorado River Basin," 14.

48. Glen MacDonald, "Water Supply," 9.

49. Ibid., 8.

50. National Research Council, Committee on Scientific Bases of Colorado River Basin Water Management, *Colorado River Basin Water Management,* 132.

51. Ibid., 1, 103.

52. Thomas F. Amistead, "Severe, Extended Droughts Seen for Colorado River Basin"; National Research Council, Committee on Scientific Bases of Colorado River Basin Water Management, *Colorado River Basin Water Management,* 14.

53. Department of the Interior, Bureau of Reclamation, Upper Colorado Region, "Drought Conditions in the West."

54. National Research Council, Committee on Scientific Bases of Colorado River Basin Water Management, *Colorado River Basin Water Management,* 66.

55. Archibold and Johnson, "Arid West No Longer Waits," A1.

56. Gelt, "Basin States Consider Ways to Share," 1.

57. Anne Minard, "Norton Rules on Water-Sharing," *Arizona Daily Star,* May 3, 2005, A5.

58. National Research Council, Committee on Scientific Bases of Colorado River Basin Water Management, *Colorado River Basin Water Management,* 9.

59. Randal C. Archibold, "Western States Agree to Water-Sharing Pact," *New York Times,* December 10, 2007, A18.

60. "Drop 2: End-of-the-Line Reservoir Salvages Colorado River Water," 1–2.

61. Jenkins, "The Efficiency Paradox," 9.

62. Quoted in Johnson and Murphy, "Drought Settles."

63. National Research Council, Committee on Scientific Bases of Colorado River Basin Water Management, *Colorado River Basin Water Management,* 71, 9 (quote), 59.

64. Glennon and Culp, "Last Green Lagoon," 944.

65. National Research Council, Committee on Scientific Bases of Colorado River Basin Water Management, *Colorado River Basin Water Management,* 9.

66. Jenkins, "The Efficiency Paradox," 13.

Bibliography

MANUSCRIPT COLLECTIONS

Bard, Thomas R. Collection. Huntington Library, San Marino, Calif.

Department of the Interior, Bureau of Indian Affairs. Land Titles and Records Office. Albuquerque.

Department of the Interior, Bureau of Land Management. Historical Indexes. Yuma, Ariz.

———. Tucson District Land Office Historical Indexes. Microfilm. Phoenix.

———. Tucson District Land Office Tract Books. Microfilm. Phoenix.

"Irrigation Data Long-Range Program, Yuma." January 29, 1944. Manuscript. Irrigation and Water Rights Case Files, 1911–1957, Record Group 75. National Archives, Pacific Southwest Region.

Lewis, Perley M. Manuscript Collection. Department of Archives and Manuscripts, Arizona State University.

Lippincott, Joseph B. "United States Bureau of Reclamation, Yuma Project Reports: Estimates and Plans, 1904." Lipp. 120-1. Water Resources Center Archives, University of California–Berkeley.

Mead, Elwood, and Burton P. Fleming. "Proceedings, Local Board of Review, Yuma Project, Arizona-California: Part I, California Lands." August 19–24, 1915. Manuscript. Mead 37, Water Resources Center Archives, University of California–Berkeley.

———. "Report of the Local Board of Review, U.S. Reclamation Service, Yuma Project, Arizona-California." January 24, 1916. Manuscript. Mead 38, Water Resources Center Archives, University of California–Berkeley.

National Archives, Pacific Southwest Region. Record Group [RG] 21: Records of the District Courts of the United States for the District of Arizona.

———. Record Group [RG] 75: Records of the Bureau of Indian Affairs, Fort Yuma Agency.

National Archives, Washington, D.C. Record Group [RG] 115: Records of the Bureau of Reclamation, General Administrative and Project Records, 1902–1919, Yuma Project.

Pardee, George Cooper. Correspondence and Papers, 1890–1941. Bancroft Library, University of California–Berkeley.

Preston, Porter J., and Charles A. Engle. "Report of Advisors on Irrigation on Indian Reservations." Vol. 1. Washington, D.C.: Department of the Interior, 1928. Microfiche. Native American Legal Materials Collection. Title 2967.

Trippel, Eugene J. "Report of the Citizens' Executive Committee on the Irrigation and Reclamation of Arid Lands." 1889. University of Arizona Special Collections, Tucson.

Yuma, Arizona, U.S.R.S. Premier Project. Yuma: Yuma Chamber of Commerce, 1927. Lipp. 120-3, Vol. 2. Water Resources Center Archives, University of California–Berkeley.

"Yuma Irrigation Project, Arizona-California: Information Compiled by United States Reclamation Service, October 1, 1909." Mead 36-1. Water Resources Center Archives, University of California–Berkeley.

Yuma Premier Project of the U.S.R.S. San Diego: Imperial Valley and San Diego County Development Bureau, ca. 1920. Water Resources Center Archives, University of California–Berkeley.

GOVERNMENT DOCUMENTS

Arizona-California, Yuma Project. "Operation and Maintenance Report for 1913." Manuscript. Bureau of Reclamation, Yuma Area Office.

Blair, R. E. *The Work of the Yuma Reclamation Project Experiment Farm in 1918.* U.S. Department of Agriculture Circular 75. Washington, D.C.: U.S. Government Printing Office, 1920.

Department of the Interior. *Fifth Annual Report of the Reclamation Service, 1906.* 59th Cong., 2d sess., House Doc. 204. Washington, D.C.: U.S. Government Printing Office, 1907.

———. *Tenth Annual Report of the Reclamation Service, 1910–1911.* Washington, D.C.: U.S. Government Printing Office, 1912.

———. *Eleventh Annual Report of the Reclamation Service, 1911–1912.* Washington, D.C.: U.S. Government Printing Office, 1913.

———. *Thirteenth Annual Report of the Reclamation Service, 1913–1914.* Washington, D.C.: U.S. Government Printing Office, 1915.

———. *Fourteenth Annual Report of the Reclamation Service, 1914–1915.* Washington, D.C.: U.S. Government Printing Office, 1915.

———. *Fifteenth Annual Report of the Reclamation Service, 1915–1916.* Washington, D.C.: U.S. Government Printing Office, 1916.

———. *Sixteenth Annual Report of the Reclamation Service, 1916–1917.* Washington, D.C.: U.S. Government Printing Office, 1917.

———. *Eighteenth Annual Report of the Reclamation Service, 1918–1919.* Washington, D.C.: U.S. Government Printing Office, 1919.

———. *Nineteenth Annual Report of the Reclamation Service, 1919–1920.* Washington, D.C.: U.S. Government Printing Office, 1920.

Department of the Interior, Bureau of Reclamation, Upper Colorado Region. "Drought Conditions in the West." In *Reclamation: Managing Water in the West.* http://usbr.gov/uc/feature/drought.html#current.

———. *Reclamation Project Data.* Washington, D.C.: U.S. Government Printing Office, 1961.

————. *Twenty-third Annual Report of the Bureau of Reclamation.* Washington, D.C.: U.S. Government Printing Office, 1924.

————. *Yuma Federal Reclamation Project, Arizona-California.* Washington, D.C.: U.S. Government Printing Office, 1936.

Department of the Interior, Office of Indian Affairs. *Indian Land Tenure, Economic Status, and Population Trends.* Pt. 10 of the National Resources Board Report on Land Planning. Washington, D.C.: U.S. Government Printing Office, 1935.

Department of the Interior, U.S. Geological Survey. *First Annual Report of the Reclamation Service, From June 17 to December 1, 1902.* 57th Cong., 2d sess., House Doc. 79. Washington, D.C.: U.S. Government Printing Office, 1903.

————. *Second Annual Report of the Reclamation Service, 1902–3.* 58th Cong., 2d sess., House Doc. 41. Washington, D.C.: U.S. Government Printing Office, 1904.

————. *Third Annual Report of the Reclamation Service, 1903–4.* 58th Cong., 3d sess., House Doc. 28. Washington, D.C.: U.S. Government Printing Office, 1905.

————. *Fourth Annual Report of the Reclamation Service, 1904–5.* 59th Cong., 1st sess., House Doc. 86. Washington, D.C.: U.S. Government Printing Office, 1906.

Executive Order, July 6, 1883. 49th Cong., 2d sess., House Doc. 1, pt. 5, 531–32.

Executive Order, January 9, 1884. 49th Cong., 2d sess., House Doc. 1, pt. 5, 531.

Fessenden, R. S. "Operation and Maintenance Report, 1916, Yuma Project, Arizona-California." Manuscript. Bureau of Reclamation, Yuma Area Office.

Heintzelman, S. P. "Headquarters, Fort Yuma, California, July 15, 1853." In *Indian Affairs on the Pacific: Message From the President of the United States, Transmitting Report in Regard to Indian Affairs on the Pacific.* 34th Cong., 3d sess., House Doc. 76, 34–53. Washington, D.C.: U.S. Government Printing Office, 1857.

Holmes, J. Garnett. "Soil Survey of the Yuma Area, Arizona." In *Field Operations of the Bureau of Soils, 1902.* Fourth Report, 777–91. U.S. Department of Agriculture, Bureau of Soils. Washington, D.C.: U.S. Government Printing Office, 1903.

Holmes, J. Garnett, et al. *Soil Survey of the Yuma Area, Arizona-California.* U.S. Department of Agriculture, Bureau of Soils. Washington, D.C.: U.S. Government Printing Office, 1905.

Johnson, Alvin. "Economic Aspects of Certain Reclamation Projects." In *Economic Problems of Reclamation,* edited by Elwood Mead, U.S. Department of Interior, 1–16. Washington, D.C.: U.S. Government Printing Office, 1929.

La Rue, E. C. *Colorado River and Its Utilization.* U.S. Geological Survey Water-Supply Paper 395. Washington, D.C.: U.S. Government Printing Office, 1916.

Newell, F. H. "National Efforts at Homemaking." In *Annual Report of the Board of Regents of the Smithsonian Institution, Showing the Operations, Expenditures, and Condition of the Institution for the Year Ending June 30, 1922,* 517–31. Washington, D.C.: U.S. Government Printing Office, 1924.

————. "Progress in Reclamation of Arid Lands in the Western United States." In *Annual Report of the Board of Regents of the Smithsonian Institution Showing the Operations, Expenditures, and Condition of the Institution for the Year Ending June 30, 1910,* 169–98. Washington, D.C.: U.S. Government Printing Office, 1911.

————. "The Reclamation of the West." In *Annual Report of the Board of Regents of the Smithsonian Institution, Showing the Operations, Expenditures, and Condition of the Institution for the Year Ending June 30, 1903*, 827–41. Washington, D.C.: U.S. Government Printing Office, 1904.

Powell, John W. *Report on the Lands of the Arid Region of the United States, With a More Detailed Account of Utah*. 45th Cong., 2d sess., House Doc. 73. Washington, D.C.: U.S. Government Printing Office, 1878.

Quechan Tribe of Fort Yuma Reservation, California. Hearings Before the Subcommittee on Indian Affairs of the Committee on Interior and Insular Affairs, United States Senate, Ninety-fourth Congress, Second Session on Oversight on Quechan Land Issue, May 3 and June 24, 1976. Washington: U.S. Government Printing Office, 1976.

Ross, Clyde P. *The Lower Gila Region, Arizona: A Geographic, Geologic, and Hydrologic Reconnaissance With a Guide to Desert Watering Places*. U.S. Geological Survey Water-Supply Paper 498. Washington: U.S. Government Printing Office, 1923.

Sellew, Francis L., U.S. Reclamation Service. "Yuma Project Historical Sketch, 1902–1912." Manuscript. Bureau of Reclamation, Yuma Area Office.

Storey, Brit Allan. *The Bureau of Reclamation's Yuma Valley Railroad*. Denver: Bureau of Reclamation, 1990.

U.S. Bureau of the Census. *Twelfth Census of the United States Taken in the Year 1900: Agriculture—Part II, Crops and Irrigation*. Washington, D.C.: U.S. Government Printing Office, 1902.

U.S. Congress. House. "Agreement With Yuma Indians in California." 53d Cong., 2d sess., House Report 1145, June 22, 1894.

U.S. Congress. Senate. "Allotment of Irrigable Lands to Indians on Colorado River and Yuma Reservations." 61st Cong., 2d sess., Senate Report 583, April 22, 1910.

————. *Federal Reclamation by Irrigation*. Message From the President of the United States Transmitting a Report Submitted to the Secretary of the Interior by the Committee of Special Advisers on Reclamation. 68th Cong., 1st sess., Senate Doc. 92. Washington, D.C.: U.S. Government Printing Office, 1924.

————. "Letter From the Secretary of the Interior, Transmitting a Copy of an Agreement With the Yuma Indians, With a Report From the Commissioner of Indian Affairs and Accompanying Papers." 53d Cong., 2d sess., Senate Doc. 68, March 19, 1894.

————. *Report of the Special Committee of the United States Senate on the Irrigation and Reclamation of Arid Lands*. 51st Cong., 1st sess., Senate Report 928. Washington, D.C.: U.S. Government Printing Office, 1890.

U.S. Reclamation Service, Yuma Project. "Historical Sketch: Appendix No. 2 Covering the Period From January 1 to December 31, 1914." Manuscript. Bureau of Reclamation, Yuma Area Office.

U.S. Reclamation Service, Yuma Project, Arizona-California. "Data Prepared for Board of Engineers, U.S.A., September 1910." Manuscript. Bureau of Reclamation, Yuma Area Office.

Youngs, F. O., et al. *Soil Survey of the Yuma-Wellton Area, Arizona-California.* U.S. Department of Agriculture, Bureau of Chemistry and Soils. Washington, D.C.: U.S. Government Printing Office, 1933.

"Yuma Project O. and M. Report, 1914." Manuscript. Bureau of Reclamation, Yuma Area Office.

"Yuma Project Annual History, 1917." Manuscript. Bureau of Reclamation, Yuma Area Office.

"Yuma Project Annual Project History and O. and M. Report, 1919." Manuscript. Bureau of Reclamation, Yuma Area Office.

"Yuma Project Annual Project History and O. and M. Report, 1920." Manuscript. Bureau of Reclamation, Yuma Area Office.

"Yuma Project Annual Project History and O. and M. Report, 1921." Manuscript. Bureau of Reclamation, Yuma Area Office.

"Yuma Project Annual Project History and O. and M. Report, 1922." Manuscript. Bureau of Reclamation, Yuma Area Office.

"Yuma Project Annual Project History and O. and M. Report, 1924." Manuscript. Bureau of Reclamation, Yuma Area Office.

"Yuma Project Annual Project History and O. and M. Report, 1925." Manuscript. Bureau of Reclamation, Yuma Area Office.

"Yuma Project, Arizona-California, Annual Project History, Calendar Year 1935." Manuscript. Bureau of Reclamation, Yuma Area Office.

OTHER SOURCES

Address of Hon. A. H. Heber to the Settlers of Imperial Valley, July 25, 1904. Los Angeles: Southern California Printing, 1904.

Amistead, Thomas F. "Severe, Extended Droughts Seen for Colorado River Basin." *Engineering News Record,* March 5, 2007, 3.

Aschmann, Homer. "The Head of the Colorado Delta." In *Geography As Human Ecology: Methodology by Example,* edited by S. R. Eyre and G. R. J. Jones, 231–63. London: Edward Arnold Publishers, 1966.

"Balancing Water Needs on the Lower Colorado River: Recommendations of the Yuma Desalting Plant/Cienega de Santa Clara Workgroup, April 22, 2005." http://www.cap-az.com/docs/newfinaldocument.pdf.

Bancroft, Hubert Howe. *The Works of Hubert Howe Bancroft.* Vol. 17, *History of Arizona and New Mexico, 1530–1888.* San Francisco: History Company, Publishers, 1889.

Bee, Robert L. *Crosscurrents Along the Colorado: The Impact of Government Policy on the Quechan Indians.* Tucson: University of Arizona Press, 1981.

Berry, Wendell. *The Gift of the Good Land: Further Essays Cultural and Agricultural.* San Francisco: North Point Press, 1981.

Bradfute, Richard Wells. *The Court of Private Land Claims: The Adjudication of Spanish and Mexican Land Grant Titles, 1891–1904.* Albuquerque: University of New Mexico Press, 1975.

Burton, Lloyd. *American Indian Water Rights and the Limits of Law.* Lawrence: University Press of Kansas, 1991.

Carlson, Leonard A. *Indians, Bureaucrats, and Land: The Dawes Act and the Decline of Indian Farming.* Westport, Conn.: Greenwood Press, 1981.

Carruthers, Ian, and Colin Clark. *The Economics of Irrigation.* Liverpool: Liverpool University Press, 1981.

Castetter, Edward F., and Willis H. Bell. *Yuman Indian Agriculture: Primitive Subsistence on the Lower Colorado and Gila Rivers.* Albuquerque: University of New Mexico Press, 1951.

Corey, H. T. "Irrigation and River Control in the Colorado River Delta." Paper No. 1270. *American Society of Civil Engineers Transactions* 76 (December 1913): 1204–1453.

deBuys, William, and Joan Myers. *Salt Dreams: Land and Water in Low-Down California.* Albuquerque: University of New Mexico Press, 1999.

Doolittle, William E. *Cultivated Landscapes of Native North America.* Oxford: Oxford University Press, 2000.

"Drop 2: End-of-the-Line Reservoir Salvages Colorado River Water." *Arizona Water Resources* 16 (May–June 2008): 1–2.

Dunbier, Roger. *The Sonoran Desert: Its Geography, Economy, and People.* Tucson: University of Arizona Press, 1968.

Fiege, Mark. *Irrigated Eden: The Making of an Agricultural Landscape in the American West.* Seattle: University of Washington Press, 1999.

Forbes, Jack D. *Warriors of the Colorado: The Yumas of the Quechan Nation and Their Neighbors.* Norman: University of Oklahoma Press, 1965.

Forbes, R. H. *The River-Irrigating Waters of Arizona: Their Character and Effects.* Bulletin no. 44. Tucson: University of Arizona Agricultural Experiment Station, 1902.

Forde, C. Daryll. *Ethnography of the Yuma Indians.* University of California Publications in American Archaeology and Ethnology, Vol. 28, no. 4, 83–278. Berkeley and Los Angeles: University of California Press, 1931.

Fradkin, Philip L. *A River No More: The Colorado River and the West.* Berkeley and Los Angeles: University of California Press, 1996.

Ganoe, John T. "The Desert Land Act in Operation, 1877–1891." *Agricultural History* 10 (January 1937): 142–57.

———. "The Origin of a National Reclamation Policy." *Mississippi Valley Historical Review* 18 (June 1931): 34–52.

Gelt, Joe. "Basin States Consider Ways to Share Colorado River Shortages." *Arizona Water Resource* 13 (July–August 2004): 1–2, 8, 12.

———. "Opposing Sides Find Common Ground in Yuma Desalter Controversy." *Arizona Water Resource* 13 (May–June 2005): 1–2, 12.

Glennon, Robert Jerome, and Peter W. Culp. "The Last Green Lagoon: How and Why the Bush Administration Should Save the Colorado River Delta." *Ecology Law Quarterly* 28, no. 4 (2002): 903–92.

Greenwald, Emily. *Reconfiguring the Reservation: The Nez Perces, Jicarilla Apaches, and the Dawes Act.* Albuquerque: University of New Mexico Press. 2002.

Griffen, Augustus. "Land Settlement of Irrigation Projects, With Discussion." Paper no. 1610. *American Society of Civil Engineers Transactions* 90 (June 1927): 750–72.

Grunsky, C. E. "The Lower Colorado River and the Salton Basin." Paper no. 1051. *American Society of Civil Engineers Transactions* 59 (December 1907): 1–62.

Hall, Sharlot M. "The Problem of the Colorado River." *Out West* 25 (October 1906): 305–32.

Harrington, John P. "A Yuma Account of Origins." *Journal of American Folklore* 21 (1908): 324–48.

Hinton, Richard J. *The Hand-Book to Arizona: Its Resources, Towns, Mines, Ruins, and Scenery.* San Francisco: Payot, Upham, 1878.

Holt, L. M. "The Reclamation Service and the Imperial Valley." *Overland Monthly,* 2d ser., 51 (January 1908): 70–75.

Hoxie, Frederick E. *A Final Promise: The Campaign to Assimilate the Indians, 1880–1920.* Lincoln: University of Nebraska Press, 1984.

Hundley, Norris, Jr. *Dividing the Waters: A Century of Controversy Between the United States and Mexico.* Berkeley and Los Angeles: University of California Press, 1966.

———. *Water and the West: The Colorado River Compact and the Politics of Water in the American West.* Berkeley and Los Angeles: University of California Press, 1975.

Jenkins, Matt. "The Efficiency Paradox." *High Country News* 39 (February 25, 2007): 8–13.

Kershner, Frederick, Jr. "George Chaffey and the Irrigation Frontier." *Agricultural History* 27 (1953): 115–22.

Knack, Martha C., and Omer C. Stewart. *As Long As the River Shall Run: An Ethnohistory of Pyramid Lake Indian Reservation.* Berkeley and Los Angeles: University of California Press, 1984.

Kniffen, Fred B. *Lower California Studies, III: The Primitive Cultural Landscape of the Colorado Delta.* University of California Publications in Geography, vol. 5, no. 2, 43–66. Berkeley and Los Angeles: University of California Press, 1932.

———. *Lower California Studies, IV: The Natural Landscape of the Colorado Delta.* University of California Publications in Geography, vol. 5, no. 4, 149–244. Berkeley and Los Angeles: University of California Press, 1932.

Kroeber, A. L. *Cultural and Natural Areas of Native North America.* University of California Publications in American Archaeology and Ethnology, vol. 38. Berkeley and Los Angeles: University of California Press, 1939.

Lampen, Dorothy. *Economic and Social Aspects of Federal Reclamation.* Johns Hopkins University Studies in Historical and Political Science. Baltimore: Johns Hopkins University Press, 1930.

Langston, Nancy. *Where Land and Water Meet: A Western Landscape Transformed.* Seattle: University of Washington Press, 2003.

Leibhardt, Barbara. "Allotment Policy in an Incongruous Legal System: The Yakima Indian Nation As a Case Study, 1887–1934." *Agricultural History* 66 (Fall 1991): 78–103.

Leopold, Aldo. *A Sand County Almanac — With Essays on Conservation From Round River*. New York: Ballantine Books, 1970.

Lewis, David Rich. *Neither Wolf nor Dog: American Indians, Environment, and Agrarian Change*. New York: Oxford University Press, 1994.

Lippincott, J. B. "The Reclamation Service in California." *Forestry and Irrigation* 10 (April 1904): 162–69.

Mabry, Jonathan B., and David A. Cleveland. "The Relevance of Indigenous Irrigation: A Comparative Analysis of Sustainability." In *Canals and Communities: Small-Scale Irrigation Systems*, edited by Jonathan B. Mabry, 227–60. Tucson: University of Arizona Press, 1996.

MacDonald, Glen. "Water Supply." In *Southern California Environmental Report Card*, edited by Ann E. Carson, 4–11. Los Angeles: UCLA Institute of the Environment, 2005.

MacDonald, Glen, Sigrid Rian, and Hugo Hildalgo. "Southern California and the 'Perfect Drought.'" In *Colorado River Basin Climate: Paleo, Present, Future*. Special Publication for the Association of California Water Agencies and Colorado River Water Users Association Conferences, 50–57, November 2005.

McCool, Daniel. *Command of the Waters: Iron Triangles, Federal Water Development, and Indian Water*. Berkeley and Los Angeles: University of California Press, 1987.

McGuire, Thomas R., William B. Lord, and Mary G. Wallace, eds. *Indian Water in the New West*. Tucson: University of Arizona Press, 1993.

Mead, Elwood. "Present Policy of the United States Bureau of Reclamation Regarding Land Settlement, With Discussion." Paper no. 1609. *American Society of Civil Engineers Transactions* 90 (June 1927): 730–49.

Meriam, Lewis. *The Problem of Indian Administration*. Baltimore: Johns Hopkins University Press, 1928.

National Research Council, Committee on Scientific Bases of Colorado River Basin Water Management. *Colorado River Basin Water Management: Evaluating and Adjusting to Hydroclimatic Variability*. Washington, D.C.: National Academies Press, 2007.

Newell, Frederick Haynes. *Irrigation Management: The Operation, Maintenance, and Betterment of Works for Bringing Water to Agricultural Lands*. New York: D. Appleton, 1916.

Nicholas, Cora Savant. "The History of Yuma Valley and Mesa With Special Emphasis on the City of Yuma, Arizona." Master's thesis, University of Southern California, 1947.

Pisani, Donald J. "Federal Reclamation and the American West in the Twentieth Century." *Agricultural History* 77 (Summer 2003): 391–419.

―――. *From the Family Farm to Agribusiness: The Irrigation Crusade in California and the West, 1850–1931*. Berkeley and Los Angeles: University of California Press, 1984.

―――. "Irrigation, Water Rights, and the Betrayal of Indian Allotment." *Environmental Review* 10 (Fall 1986): 157–76.

―――. "Reclamation and Social Engineering in the Progressive Era." *Agricultural History* 57 (January 1983): 46–63.

―――. *To Reclaim a Divided West: Water, Law, and Public Policy, 1848–1902*. Albuquerque: University of New Mexico Press, 1992.

―――. *Water and the American Government: The Reclamation Bureau, National Water Policy, and the West, 1902–1935*. Berkeley and Los Angeles: University of California Press, 2002.

Pryde, Philip R. "The Southern California Water Transfers and the Salton Sea." *Pacifica* (Fall 2002): 1, 6–8.

Reisner, Marc. *Cadillac Desert: The American West and Its Disappearing Water.* New York: Penguin Books, 1987.

Robinson, Michael C. *Water for the West: The Bureau of Reclamation, 1902–1977.* Chicago: Public Works Historical Society, 1979.

Rogers, Malcolm J. "An Outline of Yuman Prehistory." *Southwestern Journal of Anthropology* 1 (Summer 1945): 167–98.

Sauder, Robert A. *The Lost Frontier: Water Diversion in the Growth and Destruction of Owens Valley Agriculture.* Tucson: University of Arizona Press, 1994.

―――. "Patenting an Arid Frontier: Use and Abuse of the Public Land Laws in Owens Valley, California." *Annals of the Association of American Geographers* 79 (December 1989): 544–69.

Schonfeld, Robert G. "The Early Development of California's Imperial Valley." *Historical Society of Southern California* 50 (1968): 279–307, 395–426.

Shurts, John. *Indian Reserved Water Rights: The Winters Doctrine in Its Social and Legal Context, 1880s–1930s*. Norman: University of Oklahoma Press, 2000.

Smythe, William E. "An International Wedding." *Sunset* 5 (October 1900): 286–300.

Stanley, Raymond W. "Political Geography of the Yuma Border District." 2 vols. Ph.D. diss., University of California–Los Angeles, 1954.

Sutton, Imre. *Indian Land Tenure: Bibliographical Essays and a Guide to the Literature.* New York: Clearwater Publishing, 1975.

Sykes, Godfrey. "The Camino del Diablo: With Notes on a Journey in 1925." *Geographical Review* 17 (January 1927): 62–74.

―――. *The Colorado Delta.* American Geographical Society, Special Publications, no. 19. New York: Carnegie Institution of Washington and American Geographical Society, 1937.

Teele, R. P. *The Economics of Land Reclamation in the United States.* Chicago: A. W. Shaw, 1927.

Tomlinson, F. L. "Land Reclamation and Settlement in the United States." *International Review of Agricultural Economics* 4 (1926): 225–72.

Trafzer, Clifford E. *Yuma: Frontier Crossing of the Far Southwest.* Wichita: Western Heritage Books, 1980.

Trippel, Eugene J. "The Yuma Indians." Pts. 1 and 2. *Overland Monthly,* 2d ser., 13 (June 1889): 561–84; 14 (July 1889): 1–11.

The Unfriendly Attitude of the United States Government Towards the Yuma Valley, Arizona. Yuma: Yuma Valley Consolidated Water Users Association, 1907.

United States v. Coe. May 23, 1898. In *The Supreme Court Reporter,* 18:745–53. St. Paul: West Publishing, 1899.

Ward, Evan R. *Border Oasis: Water and the Political Ecology of the Colorado River Delta, 1940–1975.* Tucson: University of Arizona Press, 2003.

Warne, William E. "Land Speculation." *Reclamation Era* 33 (August 1947): 176–90.

Widstoe, John A. "History and Problems of Irrigation Development in the West, With Discussion." Paper no. 1607. *American Society of Civil Engineers Transactions* 90 (June 1927): 680–86.

Woodhouse, Connie A., Stephen T. Gray, and David M. Meko. "Updated Streamflow Reconstructions for the Upper Colorado Basin." *Water Resources Research* 42 (May 2006): 1–16.

Worster, Donald. *Rivers of Empire: Water, Aridity, and the Growth of the American West.* New York: Pantheon Books, 1985.

———. "Thinking Like a River." In *Meeting the Expectations of the Land: Essays in Sustainable Agriculture and Stewardship,* edited by Wes Jackson, Wendell Berry, and Bruce Coleman, 56–67. San Francisco: North Point Press, 1984.

Wycoff, William. "Understanding Western Places: The Historical Geographer's View." In *Western Places, American Myths: How We Think About the West,* edited by Gary J. Hausladen, 21–56. Reno: University of Nevada Press, 2003.

"Yuma Desalter Attracts International Interest." *Arizona Water Resource* 15 (July–August 2007): 5.

Index

Italic page numbers refer to illustrations, and boldface page numbers refer to maps